The Survival of Coding Genes inside the
Jungle of the Human Genome.

Illustrations by Frank Vollaro Jr.

Maya Publications
P.O. Box 200-234
Bay Ridge Station
Brooklyn, NY 11220

For further information or questions, please write to the above address.

The information in this book is not intended to be used for medical or psychiatric diagnosis of any person. The illustrations included are approximations of the subject matter for ease of understanding.

Table of Contents

69- Covalent bonds
70- Notes from my teacher
71- DNA replication process
72- Your chemistry, DNA
73- Tiny particles inside an atom
74- Ligaze protein
75- TV shows and DNA finger printing
76- Electrophore technology
77- Recombinant DNA
78- The laboratory workhorse, a bacterium
79- My noxious friend and tenant, E. Coli
80- A research hiatus
81- Protein- enzymes functions
82- Virus cannot multiply alone
83- Anxiety
84- An animal cell and its multiple organelles
85- The double helix structure for you to enjoy

February 28, 1953

The date was February 28, 1953; it was not news from the Korean War that excited James D. Watson brain cells. The day before, chemist J. Donahue from California Institute of Technology made a significant correction to the nuclei acid textbook Watson was relying on to construct a structural model for the DNA macro-molecule. How many chains would the structure is composed of: two, three or more? Where would the basis adenine, thymine, cytosine and guanine go? If placed on the outside hanging from the DNA backbone, could it come together to replicate itself creating a genetic code for all animals and plants? How could the basis A, T, C. and G, be bonded to each other and pass on the genetic message to future generations? Which basis attract, repel each other? How do we go around testing our hypothesis? The question whether protein or DNA was the hereditary material had been scientifically established at Cold Spring Harbor in New York by Alfred Hershey and Martha Chase. American and British scientists seemed to have been engaged in a competing race to discover the structure of the book of life. How was this book written?

Was it written in a code or randomly come together? Was the code a chemical or an electric one?

There were brilliant and outstanding chemist, physicist and biologist on both sides of the Atlantic Ocean very much interested in covering the first page of scientific journals and major newspapers around the world. In the U.S. A. we had world re-known, twice Nobel Prize winner chemist, Linus Pauling. He was very much engaged in the race to bring to America the Nobel trophy. In England there was Rosalind Franklin, a pioneer X- ray crystallographer who was very close to discovering the double helix DNA model. Not too far from Rosalind work place was Maurice Wilkins at the Biophysics laboratory of Kings College. These two brilliant English scientists were working close to each other; but their own bodily chemistry kept them miles apart, except when Watson and Crick, seemingly, when Watson came around trying to push his double helix model.

James D. Watson, our American biologist was desperately trying to find his place among the most prominent scientists of his time. His traveling between America and Europe put him in contact with the influential scientists and scientific organizations on both sides of the Ocean.

His scientific curiosity, accessible personality and perseverance landed him in a research position at the Cavendish Laboratory at Cambridge University in the United Kingdom. Our maverick and future Nobel Prize Laureate, J.D. Watson was blessed by sharing space in the biochemistry laboratory with a WW11, twelve years his senior, physicist, Francis Crick working toward his Ph. D. in biology. Watson and Crick, an American and a Britton seemed to have each other chemistry very well lined up to each other. Francis Crick seemed to have a pleasant, happy, gregarious and easily contagious mood and personality. In many ways, he was the perfect complement to Watson. Crick not only came out with brilliant ideas of his own, but for the good of both, Crick was

7

always attentive carefully to Watson observations when listening to his colleagues or attending lectures and conferences by other scientists. When anything did not go right, it was expected, at Cavendish, Cricks` good humor to turn it into an insightful moment to celebrate with laughing jokes and a favorable spirit to move forward with their project.

On the other hand, you had two hard headed Britton each following its own research and mind dictates. Rosalind Franklin with many years of formal training and experience over both, Crick and Watson, wanted test-result on crystallography diffraction before making any premature announcement. Cause and effect as well as test- result were her favorite criteria for claiming a job well done. The other key player in the double helix was physicist Maurice Wilkins whose past experience included working on the Manhattan Project. Whether he met and shared his work with Albert Einstein and Robert Oppenheimer, we do not know. However, we assume that Wilkins was a good team player. Despite professional intrigues, he communicated with R. Franklin and cooperated with Watson and Crick. As already mentioned, we in America had a super star chemist, L Pauling at Caltech dictating how chemistry should be read and practice. Hardly anyone would challenge his arguments as he had discovered the exact arrangement or sequence on how amino-acids fold up into proteins, the building blocks of our body. However, on the DNA double helix structure, he missed the point by a very small fraction, but still a fraction. Otherwise, he would have been a three times Nobel Laureate. Wilkins, the physicist may have much in common with Rose Franklin on research principles, but socially, they must have been worlds apart. As J. D. Watson has put it, when the Lab. day was over, Rosalind dropped her working apron-gown and put on a stylish and distinctive evening dress and mingles with London's elite society. We add that it has never been proved that we are all equal or that we enjoy eating the same apple twice.

However, it seems that neither Watson nor Crick would be intimidated by occasional road blocks impeding or delaying their search for the DNA molecule structure. United Kingdom Cambridge was a gold mine of information. Both complied with their superiors' orders to work on proteins at the expense of the double helix model. While following orders, working on proteins, their ultimate goal, the DNA, molecule structure never left their brains. Watson's eyes got caught of a scientific paper reporting that his much look- for DNA molecule bases: adenine and thymine as well as guanine and cytosine appeared in roughly equal amount. Watson brain was correct grasping this seemingly chemical balance in his golden molecule. But, it was his ever listening friend and colleague, Francis Crick that foresaw the beginning for the solution of the problem, the chemical attraction of each base pair toward each other. Adenine would attach itself only to thymine and guanine to cytosine. Two chains each holding opposite bases that chemically attract each other could explain the structural model of the double helix and subsequent duplication. Rosalind Franklin X-ray photos of the molecule were a decisive factor achieving this historic moment in human biology. On February 28, 1953, the key features of the DNA model all fell into place. The two chains were held together by strong hydrogen bonds between adenine-thymine and guanine-cytosine base pairs.[1] It all fell in place almost as Rosalind had seen it during her visit to Watson and Crick laboratory room. Subsequently, they enjoyed the Nobel Prize; unfortunately, Rosalind had died of cancer before the award was given. No Nobel Award is given posthumous.

April 25, 1953

On April 25, 1953, the world prestigious science journal, Nature, published a one page article with the title:

---

[1] -James D. Watson DNA, Alfred A. Knopp, New York, 2003, p. 52

Molecular Structure of Nucleic Acids- A structure for Deoxyribose Nucleic Acid, signed, J.D. Watson, F.H.C. Crick

There it was, a single page showing the drawing of this double helix with its sugar-phosphate backbone in the outside in a twisted fashion holding inside the nucleic acids. How tempting it was. Was it possible that I had the book of life in front of me? The authors had written, almost at the end of the article: "It has not escaped our notice that the specific pairing we have postulated suggests a possible copying mechanism for the genetic material." Is this a possible mechanism for copying the genetic material? Did it mean that there was a genetic code for me to find out how I was made? The basic elements comprising my body, including protein structure, was already shown to us by L. Pauling. Life, as it had been taught for centuries, took a different turn with this single page. From now on, the main focus of Biology was to understand the working dynamics of the double helix and its nucleic acids forming it. That single page provoked more questions than any other discovery in the XX century, according to most biologist and scientists in general. The world had had hardly time to digest and accommodate itself to Darwin's, On the Origin of Species, not to be frightened by a code of letters, A.T.C. G. to open the alphabet of our book of life.

How did the double strands separate from each other for self-replication? How and where are proteins formed? Is the code for proteins synthesis implied in the four letters? Francis Crick had postulated the dogma: DNA—RNA –Proteins. How do it relate to each other? Is the nucleus of the cell the command post for all cellular activities? Where in the cell are the letters synthesized? How many letters are in the book of life? Watson and Crick in one page article on 4/25/53 in the journal, Nature, put biologists, chemists and physicists to do science 24/7 even in their dreams.

Protein synthesis kept scientist busy until they discovered the ribosomes as the assembly site for proteins. It was Paul Zamecnik and his team at Massachusetts General

Hospital that further elucidated for us the role of R N A in protein formation. They confirmed Crick's adaptation theory naming it transfer RNA. This molecule hauls amino-acids groups that would exactly match another RNA strand coming from the cell nucleus, dubbed messenger RNA. The year was 1960. Messenger RNA was proven to be the true template for protein synthesis. We had not only discovered the letters of life alphabet, but began to learn how the acids of life work. Three world re-known American research centers, Harvard, Caltech and Cambridge, Massachusetts had done the work. M. Messelson, F. Jacobs and S. Brenner, demonstrated that a ribosomes existing outside the cell nucleus was, in effect, a protein assembly factory. There are many inside the cell; it looks like tiny beads in line in the cytoplasm. The inside of the cell was becoming clear and different organelle's structure and function was exciting news for scientists and the general public. We were working with the human genome and we were excited with the discoveries of this new science, molecular biology. Our enthusiasm prompted some individuals to make premature announcements. The cell inner organelles, its shape and function were under constant microscopic scrutiny and testing.

How nucleic acids become amino acids

Unresolved as of yet was a nagging question, how did the nucleus acid, those repeating letters: A.T.C.G., become peptide chains. In other words, there are four letters and twenty amino-acids. Which and how many amino-acids would fit to a letter to form a peptide chain? Once more, it was Francis Crick and his colleague, Sydney Brenner in 1961 at Cambridge University that proved that the code was a codon, also known as a triplet.

Three letters for each amino-acid was the solution to the problem. However, the genetic code may come in more than one codon for each amino-acid. For example, the amino-acid tryptophan- comes in UGG codon only, while arginine and serine may come out in six different triplets or codons. Word of caution, in the DNA sequence we have thymine that binds with adenine. In RNA, it is uracil that compliments adenine. Marshall Nirenberg working at the U.S. National Institute of Health participated at an International Congress in Biology in Moscow in 1961. Nirenberg was a young scientist hardly known outside his close friends. He was allowed to speak for about ten minutes. Like any other event, you like to hear the most important personalities in the field, not an unknown neophyte. Even our J.D. Watson was chatting outside the conference room. Nirenberg reported that in his Lab, he had gotten the ribosome to pump out a simple protein known as phenylalanine that comes out in codon UUU or UUC. Once more, it was Crick that arranged for Nirenberg to speak to a hall packed with every scientist that could get to Moscow the following morning. He was later awarded the Noble Prize. Nirenberg had put everyone in the run to finish the genetic code for each amino-acid.

F. Crick dogma: DNA-RNA Proteins seemed to have been well established. From the cell nucleus, DNA is transcribed into a messenger RNA molecule that goes to the ribosome in the cytoplasm. Ribosome can be seen as tiny beads (like beads in a necklace) in the cell away from the cell nucleus. Another RNA molecule named transfer RNA delivers amino-acids to the ribosome. Each ribosome acting as an assembly factory, paste together messenger RNA and transfer RNA to form or synthesize proteins. It is helpful to remember that replication exist at the double helix un-zipping process while transcription takes place while the cell nucleus is forming single RNA strands. Another term frequently used by most scientists when talking about these processes is translation. Translation is basically a process that takes place at ribosomes locations in the cytoplasm. Each molecular strand carries a set of letters that complement each other. For example, a messenger RNA triplet CAA will recognize and chemically attract the corresponding triplet GUU in the transfer RNA. The ribosomes will be in charge of pasting together these two molecules, thus forming peptide chains, known to us as proteins. Its shape will determine function and some peptide chains are called enzymes. Enzymes are very busy catalyst agent in our body; they are always provoking chemical changes without we ever knowing anything about it. As soon as I put a mouthful of ice-cream in my mouth, enzymes begin to do its work. Enzymes do their work inside the cell as well outside around the cell. We will address the ribonucleic acid or RNA later on in this essay because strands of this acid seem to be the working horse in protein forming and research tools. The traditional Watson-Crick dogma for protein synthesis has been under review recently.

We will be going back and forth in this essay trying to put together Watson and Crick milestone scientific success story. After Watson and Crick had gotten the double helix DNA structure, let us say, approved by the majority of scientists in the field, one of the questions popping up all the times was: Where and how are the building blocks of our body (protein) synthesized? What roles do the ribonucleic acid plays in life forming? Does it compete with DNA in protein synthesis? How is the ribonucleic acid involved in the complex process of peptide chains? We already advanced you some of these answers, but at the time, it was mind bugging for all involved, specially, for Watson and Crick. It was the good humored F. Crick who once again suggested the dogma: DNA —RNA—Proteins. DNA polymerase, the enzymes that unzips the double helix, was well established; soon RNA polymerase (an enzyme) came to take its corresponding place along with DNA polymerase. The nucleus of the cell was the commanding center sending out messages and ordering cells organelles what to do, but it was the multiple-forms RNA that moved around strands to assemble proteins at ribosomes sites. The template or mold for protein synthesis, dubbed messenger RNA, came out of the nucleus, but it was transported outside the nucleus into the cytoplasm to be matched with another RNA molecule carrying an amino acid destined for a ribosomes. As we indicated above, translation take place at this level of protein synthesis; it is also the place or process where most problems take place.

Translation, whether it is chemically matching nucleic acids or language translation, it seems to create headaches for all of us. It is hard to find and match words or phrases that exactly convey the same meaning and feeling in any two languages. Translation in our work is an issue of RNA molecules. During the replication process, the DNA polymerase unzips the double base pairs chain and it will re-create itself instantaneously by adding the complementary nucleotides to each strand. The original strand of letters, for example, A for adenine, T for thymine, G for guanine and C for cytosine will be followed by more letters-nucleotides- in a sugar-phosphate backbone running in the opposite direction, thus forming a double strand. The letters or nucleotides will chemically attract each other forming strong covalent bonds. They share electrons. You may have a strand composed of T. G. A. C. G. T. T. A. G. all these bases held together or attached to a sugar-phosphate bone. In this hypothetical case, the complementary bases will be as follow: A.C. T. G. C.A.A.T.C. You have T and A bonded together followed by the rest of letters running anti-parallel to each other. This is D NA replication; during genome sequencing, if you know one strand, the accompanying strand is not hard to guess. Your genes, the genetic material inside your cells, as well as every cell in all living things in planet earth is made of DNA. Each strand of DNA consisting of many poly-nucleotides is bonded together by hydrogen bonds. Although for illustration purpose we tend to draw and describe each DNA strand in a straight and bar-like structure, but in reality, it is a three dimension structure in which each strand coils around each other in a spiral fashion. Most likely, you see the length, width of the strand, but the depth in most books description, needs your imagination to be completed. We want to clarify that the word nucleotide refers not only to the four nucleic acids bases A.T.G. C.; generally, it is used to designate the bases with their sugar- phosphate group attached to it. Our DNA is well protected inside the nucleus of the cell. It is a tightly controlled structure. It needs to be that way because it needs to protect itself from invaders inside the cell cytoplasm as well as extra-cellular intrusion and interference.

Recently, the field of genetics as well as out of the womb causal factors has pointed-out risks and threats to genes and possible diseases. Before we proceed with this essay on nucleic acids, we like to remind you that the twenty amino acids needed in our organism are polymers for the synthesis of proteins. They possess an amino acid group, namely NH_2 at one end and a carboxyl acid group, COOH, at its other end. Both groups are bonded to a single carbon atom known to us as the a-carbon. There is a side chain linked to the a-carbon altering the shape of the molecule, thus changing its chemical function.[2] There are more than twenty amino acids found in nature, but only twenty are needed in body. They repeat themselves over and over again in proteins in bacteria, plants or in a privilege civilized human being like you and me. Why nature chose only twenty amino acids to be the building blocks of protein is another question to answer in the future. Also, there are about 92 elements that commonly appear in nature, but less than a third is used by life on earth. You will enjoy our trip into our body endless action and re-actions. It is a marvelous and sophisticated chemical engine. Have you wonder how D.N.A and R.N.A paired together? How about the

---

[2] - Essential Cell Biology, Alberts et al. Garland publishing Inc. New York, 1998, p. 60.

mitochondria, an organelle outside the nucleus, were it captured by the nucleus to do its work too?

Transcription follows replication

Following replication, we have transcription. With the help of RNA polymerase, the DNA molecule transcribes-meaning, forms messenger RNA, transfer RNA, interference RNA and other ribonucleic acids strands. For the sake of clarity, when you replicate (copy) a page or paragraph of a book into a home-work assignment, most probably, it will come out without any mistakes or error. But try to translate the same page from English to French, Spanish or Chinese, and in all probability, your professor or teacher will find room for improvement. If you are translating poetry, it might prove even harder to make perfect translation. As I said earlier, translation is an issue of RNA molecules. RNA is not a faithful translator or reproductions of DNA genetic instructions. Watson and Crick and colleagues established the dogma that the chemical letters A.T.C.G. are the genetic code while the cell's work horse, the RNA is made in its likeness with a minor exception, the uracil nucleotide. RNA, the nagging acid, was loaded with the responsibility of transporting molecules to ribosomes for the synthesis of proteins. These proteins are used during the making of multicellular organisms like giraffe, elephant, turtles and human beings. We wonder how many misspelling errors are found in each of genome of these animals. Researcher Mingiao Li, a geneticist at Penn Medical School in Philadelphia conducted a research study of RNA in which blood cells from 27 different persons. On average, each person has nearly 4,000 genes with RNA containing misspelling not found in DNA, while RNA has more than 20, 000 different places in the genome. The most common misspelling changes occur in the A letter in DNA which is changed to G in the RNA strand.[3] The researcher took precautions that variables,

[3] - Science News, 12/04.2010, Vol. 178, # 12,p.17

including a virus did not invalidate or stained his work. We must not forget that the world of RNA is not only very old, but abundant. While DNA sits in the nucleus of the cell giving out orders and sending out messenger RNA for the synthesis of protein, ribonucleic acids molecules are busy transporting and delivering polymers inside the cell. We already know that without messenger RNA there would not be protein synthesis and, without RNA in the ribosomes, peptides would not be pumped out of this assembly factory. Besides, micro-RNA is being used as engineering tools to work not only inside the cell organelles, but to modify genes. Do not be exaggerated concerned by so many misspellings and changes of one letter, a base for another; there is a self-correcting system inside our macro-molecule DNA. Occasional changes of letters or bases are conducive to mutations, and mutations, without losing the basic core of the cell, made you and I beautiful and intelligent human being. Ask your mother and she will re-assure you are the most beautiful and intelligence individual in the universe.

The first micro RNA

The first micro RNA was discovered in 1993; the second one was discovered at Massachusetts General Hospital in the year 2,000 by Gary Ruvkum. In between these two discoveries, in 1998, perhaps the most versatile ribonucleic acid molecule, interference RNA was added to the family. Among other things, it can interfere and suppress or inhibit a gene from forming proteins. We would like you to remember throughout this essay the potential power of t-RNA as well as micro RNA's. The potential for use and abuse of this molecule is beginning to be understood. All genetic information encoded in the DNA of eukaryote (possessing a nucleus) cell genes is transcribed by RNA polymerases. The process of transcription is dominated by proteins, but a growing number of them are RNA.[4] Recently, a new RNA molecule dubbed e RNA

---

[4] - Nature, 09/10/2010, Vol. 461, # 7261, p. 186.

(enhancer r n a) was discovered. It belongs to non-coding RNA group. Enhancers are engaged in activating the transcription process. It is suspected that an enhancer RNA contributes to human disease and the evolution of human specific traits.[5] We need not to name for you specific human traits. What makes you different from a chimpanzee besides the color of your eyes, skin, language and intelligence to make wise decisions? Besides the problem encountered by the translation process and epigenetics, we would like to briefly mention the so called "junk DNA" also known by non- coding DNA. It has been the subject of much debate and scrutiny over its usefulness in our organism. If it does not code for proteins, what is the use of having an excess of cells that may be prone to problems and disease? Is this junk DNA a left over from our earliest evolutionary developmental process that needs to modernize or recycle? Perhaps, non-coding DNA is like glia- cells, neglected by scientists because neurons, like coding DNA for proteins, has taken first place in research laboratories. An alarming 90 percent of our genome is composed of non-coding genes. A recent study done at Stanford University comparing human genetic blueprint genome with those of other animals concluded that very little of the human genome is really necessary. Andrew Sidow, the principal researcher, claims that about only 7 –seven percent of human genome is similar to the DNA of other mammals. This researcher narrows down to 225 million of the 3 billion letters (A.T.C.G.) of our DNA.[6] Not all scientists in this field concur on this estimate. During the process of transcription, when a strand of messenger RNA is being made, the junk or non-coding genes is cut out, so the messenger RNA comes out clean and safe to be transported to a ribosomes where it will be converted into a protein for your beautiful and well-shaped body.

---

[5] - Nature, 05/13, 2010, Vol. 465, p. 173.
[6] - Science News, 12/04/ 2010, Vol. 178, #12, p. 77.

Our interest in RNA

We have been interested in the RNA molecule because it takes multiple forms and in its translators' job is the subject of multiple problems that may contribute to protein dysfunction and disease. However, before we proceed on our tour of the genome and the genetic code, we are going to say some things about the DNA macromolecules. DNA has the potential for change without losing the basic components of its ancestors. It allows evolutionary processes to adapt and adjust itself to new environment, but maintains the bases of its genetic code. This genetic code has and transmits information for the purpose of creating multiple different organisms. DNA establishes amino acids sequences for protein synthesis. This great molecule created a brain capable of studying and repairing itself by protein synthesis and memory storage. (RNA molecules are prominently involved in this process). DNA molecules provide for the continuity of the species with minimum changes over a long period of time, long time may mean millions of years. If we include the Neanderthal man as part of our human ancestry lineage as I do, we have been here for a long time. Recent discoveries by Svante Paabo from the Mac Planck University in Germany, as well as discoveries at Atapuerca in Spain and southern Siberia in Russia, all point to the presence of intelligent human beings on dear mother earth long time ago.

The DNA molecule has its own membrane within the cell cytoplasm that guarantees its integrity and identity as if it were the command center of the cell. From this complex DNA molecule comes out instructions to construct a beautiful humming bird, a mouse, an elephant and a human being. It has the information to replicate itself in double helix or just simple strands of RNA. Our genetic code, the four letters: ATCG is in itself the great DNA molecule. How these letters are attracted to each other is written in DNA. This molecule keeps in its nucleus the instruction for the color of your eyes, hair and skin.

19

To some extent, it also determines how tall or short you will grow to be. Of course, environmental factors have input deciding your height and weight. A poor diet composed of simple carbohydrates without protein rich beans, grains like soy beans, green peas, lima beans, kidney beans, chick peas, and affordable tuna fish, sardines and salmon does not need to be explained any further. Likewise, throwing down your throat fried pork chops, red medium cooked steaks seasoned by artificial flavors and excess salt; you can rest assured that it would be extremely hard for you to be a healthy person at age 90.

We have become interested in the ribonucleic acid molecule not only because it transforms itself in multiple forms, but because in its interference form can inhibit the formation of our own bodily tissues, including brain cells. The micro RNA hair pin molecule is also another relatively new discovery. Thanks to its small size, this molecule has been used to deliver research and therapeutic tools inside the cell. It can be used to engineer gene modification in a brain disease or any other bodily pathology. It has been used in lower animal models. Time, dedication and, of course, funding, will decide the future of all these new technologies. Micro RNA can be used to target genes that at a certain age or bodily development provoke a dysfunction or a specific disease. These diminutive molecules can take its place in amino acids chains and alter the completion and function of a faulty protein or gene. During the blastocyst phase of embryonic growth, it can be used to silence pathological cells. Gene silencing through RNA interference and similar micro RNAs has become the tool of choice for analysis of gene function, diseases and drug discovery. Among the smallest of these molecules besides the micro RNA, we have short interfering and hairpin screening tools which are first choice among genome researchers.

Fighting Diseases

Our next challenge to Watson and Crick book of life is how the genetic code can be translated into fighting diseases and improving human health in general? From studies of mummies and animal fossils we have learned that some diseases have been haunting us for thousands of years. The study of nucleic acids- the base pair of letters- in our genome can be used to point out mutations, missing segments of letters, misspellings and repeating non coding genes. Why has it been so difficult to fight cancer, schizophrenia, Alzheimer and other degenerative diseases in our organism? We have engineered animal models for laboratory work like mouse, worms, bacteria and fruit fly, among others. It has been a challenge for all of us that would like to see human suffering alleviate with newly found discoveries, mainly after WW11. Privately funded laboratory researchers, pharmaceutical companies, State and Federal agencies joined in the effort to find a solution as soon as possible. During these intervening years much funding was diverted to space exploration in competition with the Soviet Union. The exploration of human somatic and brain cells was lagging behind. We did not have the technology-the working tools- to observe, catalogue, categorize and analyze the behavior and role of each DNA molecule. The same work had to be done with the ribonucleic acid molecules. Whatever little work done, it was done by hands! Although the double helix and both nucleic acids were known in general, the dynamics, physiology and chemistry of the brain was left to Sigmund Freud, C.C. Jung, theologians and poets to speculate about it. For two thousand years, our cultural and religious heritage from Greek and Roman philosophers had speculated about our illness, especially, mental diseases. The old Greek scientific knowledge was buried under superstition and religious fanatics during the middle ages. It was not until the Renaissance that scholars from Byzantium introduced it to Western Europe.

The mouse brain tissue

The slicing of a mouse brain tissue, observing how a cell grows in a Petri dish or a virus multiplying in a bacterium environment did not receive the necessary support from the public. For gene sequence history lovers that enjoy engaging themselves in mind travel, we like to remind you that by 1985, the longest genome that had been sequenced was a virus of 172,282 base pairs. The human genome has a little over three billions. Do you understand what we mean? We were sleeping, and sleeping soundly. That was 32 years after Watson and Crick had given us the genetic code of life. Three worldwide known leaders and researchers took over the audacious if not gargantuan proposal: SEQUENCING THE HUMAN GENOME. We were still guessing or speculating about how many genes are there in the genome. The three visionary leaders we are referring to are, James D. Watson, Francis Collins and Craig Venter of the Celera Genomics. As late as 1988, James D. Watson was named associate director of the Office of Genome Research, part of U.S.A. National Institute of Health. He was in charge of the genome project until 1992. Francis Collins took over from Watson until its sequence was completed in 2003. Craig Venter published his human genome sequence in the journal Science while the N.I. H genome was published in Nature, a British journal. By the time Watson, Collins and Venter took upon themselves the sequencing of the human DNA. P.C.R. the tool that multiplies DNA fragments for analysis was available to them. Making use of restriction enzyme to splice or cut DNA strands at target sites became very useful tools not only for reading the sequence, but to mark genes involved in inheritance diseases. Some of the earliest molecular biologist and scientists working with DNA structure and its content claimed that genes made up only three percent of DNA in human cells. That claim was as late as 1995. Today, we call non-coding DNA introns while coding DNA is dubbed exons. As more sophisticate research tools are developed, the so called junk DNA become under meticulous scrutiny and analysis. These non- coding genes are revealing their usefulness. The race to be the first one to sequence and read the book of life was centered in America with minor international contributions. However, as government and scientists around the world became aware of its potential benefit not only as a wonder tool fighting disease; but also, as an agricultural aid improving food supply to the world population, network were established for joint research projects. Seeds can be engineer to improve harvesting seeds, accelerate animal and chicken growth. Genetic engineering can be the tool per- excellence to feed hungry people around the world where climate and weather disasters make it difficult to grow enough food for daily consumption.

The GWA

For this purpose, Genome Wide Association- G.W.A- studies have engaged hundreds of individuals, both researchers and patients, identifying disease markers in their DNA. Deciphering the book of life language has arisen- wake up- the mind of many brilliant and concerned persons on planet earth. Using the advances of Restriction Enzymes and Recombinant DNA, we have moved forward in our attempt to bring relief to the suffering and hopeless individuals worldwide. We are using genes to help patients, as in gene therapy. We have been using harnessed virus as a vector to deliver therapeutic tools to target areas. The usefulness of the human genome sequence benefits is not in question. Many individuals have hailed it as the greatest human achievement ever made in human history. Of course, there are alarmists that have religious, philosophical or superstition claims that interpret man's advances a challenge to their world view. We are going to enter two quotations from two different professors addressing Craig Venter latest achievement with a synthetic bacterium. "A prosthetic genome hasten the day when life forms can be made entirely from non-living materials. Another professor wrote: Venter's achievement would seem to extinguish the argument that life requires a special force or power to exist. This makes it one of the most important scientific achievements in the history of mankind." [7] Following, I would like you to ponder in the quotation by another professor of biomedical engineering that appeared on page 424 of the same journal. "The microorganism reported by Venter team is synthetic in the sense that its DNA is synthesized, not that a new life has been created. Its genome is a stitched copy of the DNA of an organism that existed in nature, with a few small tweaks thrown in." You have the opinion of three different professors, all reading the same article in the same journal. One praises it, another puts down while another sees it as an assault to almost we believe in. When C. Venter appeared in T.V. announcing the creation of a synthetic bacterium, we do not recall him claiming the forming of new life. We understood it as a brilliant genetic engineering accomplishment of mankind never seen or done before. He had opened the door for a new approach to make genes and proteins work for us at our own level of development. We have many variant genes that seem to be idle or not performing at our own level of self-development. Venter has proven that we have grown up, he has proven our humanity; we can take charge of our own organism. He just finished sequencing his own genome, practically a few minutes ago in our life history.

C. Venter challenged HGP

C, Venter had challenged the Federal Government sponsored Human Genome Project and came out a big winner. His success relied upon improving existing technology to assemble and improve on nature already existing life structure. Life itself is dependent on little beats of elements coming together and functioning as a separate unit. These units of life

---

[7] - Nature, 05/272010, Vol. 465, #7297, p. 422-423

are around us, and the universe is full of it. C. Venter concept of throwing in little pieces of life elements (genetic engineering) into the cell was around for almost four decades. We are referring to recombinant DNA pioneer, Herb Boyer and Stanley Cohen work. Without hesitation, we can call them the first and most accomplished genetic engineers. The beauty of C. Venter team creature, the synthetic bacterium, is its ability to engineer organic material in the universe (our planet is part of, and the universe itself) to assist and improve human evolution. When we first cut and isolated DNA molecule making use of restriction enzymes, thus putting our fingers unto genes, we continued our game with our alphabet of life throwing in the ligaze enzyme, and bingo, it glue together and a plasmid came to being-exist as another tool. These two new tools of research and therapy: restriction enzyme and recombinant DNA took over the front pages of scientific journals, newspapers and newscast on T.V.

When we were cloning DNA sequence, cutting with restriction enzymes and linking it together with a DNA enzyme named ligaze, we were throwing in "little tweaks" in the genome cells. We were using the bacteria cell as the working horse as it divides over and over again in culture. Boyer and Cohen were using an existing micro-organism for the benefit of a multi cellular organism. We can argue that Boyer, Cohen, Venter and other scientists are just accelerating and improving on the evolutionary work begun many millions of years ago by eukaryote organisms. The evolution of life on earth has had multiple twists and turns not fully clear to most of us. One of those twists must have been when the RNA molecule made advances to another powerful and interesting molecule, the deoxyribonucleic acid molecule, DNA.

Francis Crick and RNA

Our friend and brilliant scientist, Francis Crick, according J.D. Watson, as early as 1968, suggested that RNA must have been the first genetic molecule…besides acting as a template, might also act as an enzyme catalyzing its own self replication."[8] This insight of Crick on the ribonucleic acid when the double helix and its nucleotide were still trying their rightful place in our brains was awesome. This fellow seems to pull his ideas out of his jacket rear pockets, and most of ideas were proven right. Even on his way to the hospital before his death, he was doing science in the ambulance, according to some of his biographers. We are grateful we have guys like him working for humanity.

A team of researchers from one of the most prestigious Universities in the U.S.A. wrote: "The reality of R.N.A., an R.N.A. dominated stage in the early evolution of life prior to the evolution of coded protein synthesis, has been firmly established by recent structural studies of the ribosomes.[9] We must point out that the protein assembly factory, the ribosomes; one of its components is ribonucleic acid. This acid in its multiple molecules is not only an effective working horse, but in the ribosomes, it is part of the machine that builds very complex organisms like human beings. Further elucidating cellular and life complexity on planet earth, we must attempt to answer, how did the mitochondria- the cellular energy engine- came to be part of our organism's cells? Nature has been working with and engineering our basic functional units, our cells, from the very beginning of life itself. All our basic elements beginning with carbon nitrogen, hydrogen, oxygen, potassium, sodium, etc., are part of nature. Watson, Crick, Cajal, Collins, Cohen, Boyer, Venter and all of us, is just nature in its natural evolutionary development. It seems that they are among nature most developed creature. The contribution by pioneer scientists sequencing the human genome has helped our relationship with other species. It has helped us understand

---

[8] - The R. N.A. World, 3rd edition, Cold Spring Harbor Laboratory Press, N.Y., 2006, p. XXIII.
[9] - Op. cit., p. 57.

26

ourselves by understanding our closest hominid genome, the chimpanzee. The mouse, not too far behind, is among the best research tools we have. By engineering and observing human diseases in a mouse, we can trace genes responsible for the disease and develop corrective therapeutic tools.

Current worldwide issues like climate change may also benefit from hominid genome sequencing. Anthropologist, historians and scientists of multiple orientations must have enjoyed Svante Paabo working at the Max Planck Institute in Germany latest discovery on homo- sapiens travel around our planet. He brought up new issue with out of Africa theory as part of mankind history on our planet. He posed serious questions about climate changes and its impact on plants and animals survival on earth. Neanderthal once occupied most of Europe from the Iberian Peninsula to southern Siberia in Russia. Climate change brought drastic weather changes and most of the continent was covered by heavy snow all year around. How much did glaciers contributed to Neanderthal extinction needs further study and analysis. This study relates to us in terms of demographic movement around the world and its impact on us. The sequence of human genome from different part of the world population helps understand how genetic and the environment influence each other in disease etiology. Which genetic factors are basically biological and which are environmental factors. Through worldwide genome studies we can identify the most common variants from samples in China, Japan, India, Russia, Europe and U.S.A. For example, does autism appears in the same gene and chromosome in children in any of those countries? Biomedical technology developed in the U.S.A. using multiple RNA molecules tools can be used to bring East and West together fighting diseases and plagues in animals and plants. It seems we trade with every country in this planet, we eat their food stuff and handle many of their made products that may provoke spontaneous diseases.

After the HGP completion in 2003, what about?

Following the completion of the Human Genome Project in 2003, many surprises about genes structure in the chromosome have come to light. Large segments of non-coding genes appear far away from genes associated with a disease, but it looks that it is linked to the disease. In some cases, genes variants common to autism have been linked to brain diseases like schizophrenia and manic depressive illness, also known as bipolar depression. Autism, like schizophrenia is a multiple symptoms brain disorder. Multiple genes and environmental causal factors may be involved in triggering its behavior. A group of international researchers studied nearly a thousand autistic persons and compared it with an equal number of healthy individuals. They found dozens of genes involved in autistic disorder. Interesting and rightfully surprised was the fact that most of the variants in autistic disease were missing segments or duplicated segments of DNA. Some of the genetic changes were inherited while other has to be attributed to parents or the child himself (epigenetic).[10] Outstanding symptoms of autistic children are their severe social limitations or difficulties, besides language problems and repetitive behaviors. We will include the following research finding of 2006 because it covers a large number of families with autistic members. "The gene dubbed MET regulates production of a protein that influences cell proliferation in various parts of the body. This gene lies on a stretch of chromosome 7 that other researchers had linked to autism. Neurologist Pat Levitt from Vanderbilt University, the team leader, consulted database from 204 families with one or more children with autism. Another group of 539 families was include in the study and found that the link between the Met variant and autism appeared primarily in families with two or more affected children. Laboratory tests showed that this MET form lowers the gene's activity and reduces its production of proteins that

---

[10] -Science News, 10/23/2010, p.18-21

binds to various tissues. Daniel H. Geschwind from the University of California at Los Angeles praised the report as the first time someone had identified a candidate gene for autism; however, he cautioned that Met could be the tip of the genetic iceberg. [11] A few years have passed ever since Pat Levitt conducted and published this report. As quote number 10 indicates, more recent research done by scientists attempting to untangle the genetic roots of autism, have found dozens of genes involved in this brain disease. But, we emphasize once more, there is more than just genes implicated in this disease. Autism, like schizophrenia, is not limited to race, country, ethnicity or social status, common efforts to find a cure might not be too far away.

Autism affects mostly boys. Finding out the etiology of a genetic disease is not just pointing out gene duplication, missing segments of the genome sequence, molecular intrusion such as short nucleotide polyphormism (SNP) or a mountain of non- coding genes. Looking for rare variations in human genome is only one approach to multiple rare syndromes like schizophrenia and autism. What caused genome segments to be missing or duplicated, in the first place? What provokes SNP to appear in harmful locations? Researchers have shown that not everything is inherited. How about the embryonic phase of gestation? Parental mental and physical state at the time of conception undoubtedly contributes to embryonic, fetus, childhood and adults problems. The human embryo is subject to multiple risks and threats from inside as well as outside the mother's womb. Inside is not only a gene mutation by a letter change like A replaced by G. You have to consider epigenetic, a set of chemical alterations not found in the DNA code that partially manipulates gene expression. Among epigenetic markers we have methylation (CH_3, acetylation, phosphorylation and similar markers that may attach themselves to histones tails and alter critical transcription processes. Histones are proteins that DNA coils around it.

[11] - Science News, Vol.170, 10/21/2006, p.259.

Further complicating our job, we have micro RNA that may bind itself to a messenger RNA and suppress the synthesis of a particular protein. Suppose the protein that was not synthesized to its shape was incomplete or modified in some way and could not take its rightful place in a visual pathway in the occipital lobe? Definitely, it would put our vision in jeopardy. In this case scenario, outside environmental threats and risks has to be considered secondary to biological causal factors. A biological culprit can be found in any other vital organ of our body and not necessarily a dysfunctional gene.

DNA packing

DNA packing could become an area of extreme care susceptible to harmful epigenetic intervention. Histones are proteins responsible for the first level of packing in eukaryote (cells with a nucleus) chromatin. The complex comprising DNA and proteins (histones) that forms a chromosome is called chromatin. The amount of histones in chromatin is approximately equal to the amount of DNA. Histones have a high proportion of positively charged amino acids (lysine, arginine) and bind tightly to the negatively charged DNA, thus forming the tightly packed chromatin. This tightly packed complex is a safeguard measure against outside intruders. It can be seen as nature attempt to keep DNA uncontaminated and complete as possible. Seen under an electronic microscope, unfolded chromatin appears looking like beads of a string or necklace. For those with a religious orientation, it looks like a rosary beads. Each bead is called a nucleosome. As you may have already suspected, chromatin organization in nucleosome may influence gene expression by limiting the access of transcription proteins to DNA.[12]

In summary, chromatin is the molecular substance in all eukaryote cells that facilitates DNA packing efficiently. It

---

[12] -Biology, Neil A. Campbell, Benjamin Cummings Publishing Co. Ca. 1987, p. 372-73.

protects itself from in-necessary un-raveling, unwinding itself and exposing its nucleic acids: adenine, thymine, guanine and cytosine to risky alterations. The DNA winds up, meaning, coils itself around histones, which are proteins, which look like tiny beads. They make a good binding to each other through negative and positive charges. However, for reasons we do not understand at present time, genome development through millions of years, allowed histones to grow tiny tails coming out of its beads like structure. These tails, like any external protuberance may serve well the molecule, but the tails can be a tempting target for SNP and other intruders to anchor and provoke havoc to our life creating acid, DNA. On the other hand, histones tails may serve as anchor sites for extra- cellular transporter of nutrients for both, DNA and histones. Besides growth factors, it may also serve as listening post for communication among cells in bodily organs, and between neurons in the brain. How the cell keeps a balance between growth factors and SNPs while conserving for itself continuity-- the core genes that will keep the specie genotype and phenotype (outside appearance) with minor changes-- is another question we have not been able to answer in a satisfactory fashion.

A seemingly contradiction

It seems a contradiction for us that histones proteins around which DNA is tightly wrapped, allows tiny tails that may threaten its own structure and function. Cells are chemical factories going through multiple processes to keep its intrinsic functions. The cell has many channels in its membrane, including the energy producing dynamic molecule ATP (adenine triphosphate) that plays a very significant role in metabolic processes. Does it have a role, a position protecting the chromosome complex, in particular, the histones` tails from undesirable intruders? Multiple proteins are involved in forming the chromatin complex, including the critical

31

transcription and translation processes that are critically important for protein synthesis. We have mentioned before that multiple mistakes of different sorts take place during these two protein synthesis stages. Proteins are crucial for DNA tightly coiling-wrapping around and shape forming besides being part of ribosomes. A group of researchers interested, among other things, in DNA and histones packaging in chromatin, studied an enzyme dubbed, 1sw2. It is an ATP-dependent chromatin remodeling protein highly evolutionary conserved. They found that the positioning of thousands of nucleosomes adjacent to important regulatory sequences is controlled by 1sw2. This enzyme is able to use energy from ATP hydrolysis to override the inherent nucleosome-positioning signal of the underlying DNA. Isw2 may function generally to reposition nucleosomes on unfavorable DNA sequences. The ability of proteins such as 1sw2 to reposition nucleosomes provides a clear illustration that cellular factors activity operate to disrupt the intrinsic cues that would otherwise package the genome.[13]

We hope to have given you an idea, although incomplete, about the multiple problems we encounter just trying to identify possible loci for a disease. As mentioned earlier, it is not just carrying the genes of one parent or the other, but epigenetic and everything around us. Continuing with epigenetic, Frances Champagne, a neuroscientist at Columbia University in New York has said that DNA methylation is the most enduring of epigenetic modification. It seems to be a complicated system of DNA synthesis, packing, replication and transcription. In a base pair, guanine-cytosine, the methyl molecule CH_3 may just sit itself next to cytosine nucleotide in the double helix and become methylated. The problem is that methylated cytosine looks like another nucleotide called thymine and may be converted from cytosine to thymine. Look at this molecular scenario and think about the possible problems in protein synthesis in this DNA mismatch.

---

[13] Nature, 12/13/07, Vol. 450, #7172, p. 1031-35

Suppose this protein destiny was either the heart or the occipital lobe, our vision region in the back of our brain; in all probability, it would provoke serious problems to our eye-sight. Another area of great concerns for us is the pre-frontal cortex, the last brain region to attain maturity. It is vital important in cognition, rational thinking- along with decision making. Most recently researchers have found that the prefrontal cortex is intrinsically connected with the limbic system in decision making. A rational decision is dependent on input from both brain regions. During adolescent hood, the time when we are neither an adult nor a child, hormones release is triggered by simple stimuli, emotions arisen from the limbic system. The emotional stress and physiological demands we go through during this phase of our life development and growth is, perhaps, the most critical during our entire life. Stress may be provoked by many things inside and outside us. Unconsciously, we may be under continuous stress caused by childhood painful experience that pre-dispose a person to multiple illnesses later in life. Stress may have come from parental disputes and illness, natural disasters like earthquake, tsunamis, fires, poor nutrition and civil strife. A poor diet may contribute to inadequate consumption of amino acids food; amino acids are the building blocks for proteins and the latter are our body building blocks.

DNA Methylation

Before we proceed to elaborate another theme, we would like to say a few more things about methylation. DNA methylation is a modification that controls gene expression. This is a very important and strong statement; it says that control gene expression. If it controls gene expression, consequently, the whole process of protein synthesis comes under its influence. Methylation, which is a molecule, composed of one carbon atom and three hydrogen atoms bonded to DNA. Besides, DNA methylation contributes to mammalian development, aging and some diseases. In humans and in mice, our lovely research tools, the addition of a methyl group to cytosine within a cytosine- guanine dinucleotide is catalyzed by DNA methyl-transferase and enzymes DNMT3A, DNMT3B or DNMT1. The latter is the maintenance methylase because it adds a methyl group primarily to double strand DNA that is already methylated on one strand. I promise not to repeat it again, but the cytosine-guanine dinucleotide the researcher is referring to are the letters T and G that we talked about in the double helix. When a methyl group is added to the DNA bases, it will modify the sequences, for example GAAT. The methyl molecule CH_3 attachment to the letters just mentioned will make the letters group (gene) look different and the restriction enzymes that cut when it recognizes a particular sequence will bypass the sequence GAAT. Methyl is not the only DNA marker, and they play multiple roles. Do not hesitate reading it, the next paragraph are much easier. You must keep in mind that you have brave and extremely smart cells in your body. Your DNA molecule plays around with four letters or bases, amino acids, RNA and multiple molecular transformations before a protein comes out to form part of your eyes, heart, brain or legs for you to move around. All these are very complicated processes that are taking place inside you; you can say you are doing it because nobody else is around to do it for you.

Chronic stress and your brain

The pre-frontal cortex is often associated with schizophrenia, bipolar depression, anxiety and a host of other brain disorders. It seems the cell has insurmountable problems to conquer to be able to survive. Of course, it has developed its own repair system including demethylation. J.D. David Sweatt at the University of Alabama at Birmingham doing research on methylation writes that demethylation occurs rapidly under certain conditions, such as when people experience stress.[14] Chronic stress has been demonstrated to be extremely harmful

---

[14] -Science News, 05/24/2008, Vol.173,#17, p.19

to our mental and physical health. Brain organs like the amygdala on both temporal lobes and seat of learned fear, and the hippocampus on the right and left hemisphere of our brain and site per excellence for memory and learning is seriously compromised under stress. Some researcher have found that chronic stress kill cells in both hippocampus, thus seriously compromising your ability for learning and memory. An over-stressed amygdala would not be able to respond properly and fast enough when threatened from outside your body and, you may be an easy target for a predator. While in the jungle, the amygdala saved us from becoming the lunch or dinner of a hungry tiger, lion or bitten by snakes. In big urban areas, the amygdala helps save us from human predators. Similarly, virus and bacteria otherwise dormant, when your defense system is weakened by chronic stress, may seriously threaten your health. Most of us are aware of stress situation and its harmful effect on blood pressure, heart problems as well as tension headache. We usually run to the pharmacy to get over the counter medication to get rid of pain and discomfort from chronic stress, but rarely stop to think about the harm we are doing to our body. Your entire immune system is seriously compromised under stress, particularly, chronic stress, and symptoms and diseases otherwise dormant under the watchful eye of T and B cells and an army of assistant cells. Bacteria and viruses disease provoking do not miss a chance to attack healthy cells and get us ill. One more note on the amygdala. In recent years (published in Science News, 02/26/2011, the surge of new research has expanded scientists view of the amygdala importance. It turns out the amygdala helps shape behavior in response to all sorts of stimuli, bad and good. It plays a role not only in aversion to fright, but also in pursuit of pleasure. It has neural connections with all five senses. This is a very primitive organ of your brain which assigns value to reward and adjust that value as circumstances changes. It also suggests that the amygdala plays a role in goal oriented behavior. Having connections with all senses, it is in a unique position to take immediate precautionary action when necessary and involve our commander in chief region of our brain, the prefrontal cortex, for rational decisions. We must add that connections between the amygdala and the thalamus are very, very close. If necessary, the axis composed of thalamus, hypothalamus, pituitary and adrenal gland will be pumping hormones all over your body and my body. I am not aware that I am doing it, but my heritable genes that I got from my ancestors equipped me with emergency tools for protection against unforeseeable threats.

Our body energy and wisdom

Your body and my body have in storage vast amount of energy and a treasure of accumulated knowledge and wisdom. Have you ever thought of the amount of energy stored in each atom of your body? Ever since we dropped the first atomic bomb in Japan I have been wondering about the potential power of atoms in my body. Einstein, Bohr, Oppenheimer and the like have put ideas in my brain that make me wonder how I exist in planet earth. I am made of atoms that form molecules that will become cells and ultimately organs and tissues forming my body. However, the space in atoms in my body is more than 99 percent empty with electrons traveling in orbital fashion at nearly light speed. The subatomic particles traveling at such speed inside my atoms inside my body seems to me to be a walking powerful energy warehouse. I have very little control of what is going on in this chemical and atomic machine of mine. And… you are right, how can I say that this machine is mine? My organism is continually processing external perceptions and internal chemical and atomic reactions. When I am witness to a car accident on the highway and I see people injured, specially, if the injured are children and women, I will have millions of chemical reactions taking place inside me that I am not aware of. Sadness and fear might be the first feeling registered in my brain. The thought of that I could have been involved in the accident places my amygdala in a state of alert. There is an amygdala in each temporal lobe. The amygdala will automatically alert my sympathetic nervous system for an immediate reaction if it is called for. If it is life threatening, the amygdala will either fight or flight if it does not chicken out and freezes. In the mean-time, the amygdala will connect with the thalamus that in turn will get in touch with the hypothalamus which will release a message to the pituitary gland which will release a hormone down to the adrenaline gland. The adrenaline coming from this gland will move up the body and trigger a myriad of responses through millions of chemical and electric reactions. Needless to say, my limbic system and prefrontal cortex must have been involved in all decisions in my brain, except, perhaps, the split second reaction of the amygdala in a life threatening situation. A point that I am trying to convey through this hypothetical illustration is that a visual stimuli with no matter attached to it, is capable of provoking in my organism (maybe in yours too) a series of behavioral responses in which matter and electricity were involved. Matter was converted into energy, and energy, like in neurons, is converted into matter. I do not know an iota how this transformation takes place, but rest assure, it is taking place every fraction of a second in my body. This stressful situation I just described for you compromises my health, among other things, the immune system. My heart palpitations went up, blood pressure, asthma symptoms and headache became unavoidable health threats. You may dismiss this health lesson as a trivial every day episode of modern life style. At issue is your own reaction to daily life episodes. How many stressors do we encounter in a day, a week, a month or during a year? In this biochemical scenario, your DNA, RNA and its many micro molecules are modifying the structure and function of cells, organs and tissues in your body. A hyperactive amygdala under chronic stress can be a serious health problem. An excess or deficit of a neurotransmitter in the brain can lead to a brain disease. The sooner we become aware of genes, epigenetic and environmental causal factors, the better for our health and happy life.

Through worldwide network studies, researchers are sharing genome connected information regarding diseases. It allows a researcher or research team to follow up on another person's findings and ideas. Not all researchers have access to money and technology to go to next level of investigation in their project. Some countries and institutions provide research funds only when the by-product is close at hand and has an immediate and necessary use like in disease and food production for its population needs. Long term basic research, if not ignored, is pushed aside. In the U.S.A. and Europe, basic research takes in a good sum of money. The sequencing of the human genome, restriction enzymes, recombinant DNA, gene therapy, deep brain stimulation, just to name what pop-up in my brain, are all American accomplishment. James D. Watson put together the double helix together in England, but he is an American y birth. Most recently, our scientists are exploring how to engineer genes to make proteins at the moment and location we need to have it. A new research tool dubbed ontogenetic would allow neuroscientists not just observe RNA molecules respond to DNA messages, but will empower it to manipulate neurons at will. Gero Miessenbock at Oxford University in England says: Light responsive molecules used in ontogenetic experiments uses a specific wavelength of light... a channel opens and allows positively charged ions to flow into the cell. This happens to be the neural code for on, K Deisseroth from Stanford University said. Other light responsive molecules, when tickled with the corresponding wavelength of light will allow negatively charged ions into the cell. Using a combination of the two types of molecules and different wavelength of light, researchers can flip neurons on and off at will to find out how neurons interact with their neighbors.[15]

---

[15] -Science News, 01/30.2010, Vol. 177, # 3, p.18-21.

Hopefully, this new research tool will allow us to look at neurons at the hippocampus and actually observe the neuronal behavior in learning and memory formation. Similarly, focusing on the amygdala will show you learned fear. The nucleus accumbens and the tegmental ventral area in and around the midbrain will teach us how addiction behavior is interconnected in that area of the brain. We are moving forward at great length after the discovery of the double helix and the genetic code in pursuit of our goal deciphering and making correction on gene misspelling, mutations as well as translation processes multiple mistakes.

With your permission, we are going back to SNP and missing nucleic acids letters like A T C and G on D.N. A. as it pops-up in research literature very often. It is well known that tagging-binding- a SNP such as a methyl group to DNA, generally suppress, meaning shuts down, genes next to it. We already mentioned the possible harmful effects of a missing or altered genes in protein synthesis. Rare mutations- single letters DNA changes- can contribute in a big way whether a person gets a disease or remains relatively healthy. Some variants have been connected to common disease like schizophrenia, bipolar depression and mental retardation. Determining the genetic base for any of these diseases will take time and perseverance. It is not only a matter of genetics, epigenetic and nurture, but the conformation of the disease itself. In schizophrenia you may rightfully argue that the whole brain is involved in producing multiple symptoms. It is cognitive, affective, visual, auditory, covers smell, taste, skin and other symptoms that are hard to attribute it to a single word of misspelling in the DNA alphabet. However, general wide studies with many participating researchers engaging thousands of patients worldwide, might point the right direction while pursuing a particular disease symptom pathway.

We love to repeat teaching lessons from James D. Watson. He did not stop working for us. His goal was not to leave for us the double helix, but the sequencing of our genetic code as it relates to our health. He wrote: A single base change in the DNA sequence of human beta hemoglobin gene results in the incorporation of the amino acid valine rather than glutamine acid into the protein. This simple difference causes sickle cell disease, in which red blood cells become distorted into a characteristic sickle shape.[16] Watson was mind traveling in space and time during his work at Cavendish Laboratory along with F. Crick in the early 1950s. As I was growing up I remember we used a sickle shape blade to clear weed and small bushes growing under coffee trees. In some Latin American countries, the economy is based on sugar cane, tobacco and coffee crops. Therefore, the sickle shape blade was a valuable working tool.

Michael inquisitive mind

Once more we will repeat this mutation process to satisfy our inquisitor's mind, Michael. We need three nucleotides or letters of DNA alphabet which normally has four: A.T.G and C. for binding to each other. Three letters specify to form a codon. In most cases the genetic code for each amino acid can be multiple codons composed of three different letters. For example, the amino acid leucine can be formed by six different codons while tryptophane is formed by only one codon, namely U.G.G. Scientist Vernon Ingram at Cavendish Lab. had found out that amino acid glutamine formed by codons G.A.A. or G.A.G. found at position 6 in the normal protein chain, was replaced by valine that can come in four different codons of the genetic code: G.U.C., G.U.U., G.U.A, G.U.G. in sickle cell hemoglobin. You must remember the four letters found in DNA and its difference to the RNA chain. RNA has uracil as a nucleotide, while in DNA; thymine

---

[16] - James D. Watson, op cit., 67.

is a complementary nucleotide to adenine. In RNA, it is uracil that complements adenine. You have this two great molecules working together to build tissues for your body. You have DNA polymerase that unzips DNA during the replication process. However, during the transcription process in the making or forming a messenger RNA, the RNA polymerase helps creating a single strand. Messenger RNA carries the message, and is the template for protein synthesis.

This messenger RNA that was transcribed from the DNA cell nucleus will end up at a ribosome. The sequence of the messenger RNA will be used at each ribosome to form or synthesize a new protein molecule. Another molecule, transference RNA carrying amino acids will cross the cytoplasm and anchor itself at ribosomes. At one end of this t RNA there is a set of letters, let us say, UUC that recognize and establish a bond with its opposite codon, A. A. G. in the messenger RNA, with the resulting codification for lysine. An RNA molecule serves as a template for translation of genes into proteins, another RNA molecule transfer amino acids to the ribosomes, the assembly site for protein synthesis, besides, it conforms into other useful small forms. You have also seen that multiple codons can specify for the same amino acid. This dynamic process could create problems in protein synthesis. Take, for instance, leucine with six different codons and tryptohane with only one codon; it would appear that protein folding from tryptophane only one codon would be much less mistake prone than six codon, leucine. Now consider variants like nucleotide polymers components, its availability in the cell, and the intervening stages in codon selection, availability, peptide chains and final protein folding is enough room for multiple genome mistakes and consequent disease. You may want to accuse us of picking on translators, but this RNA transfer molecule has to do a superior job not to commit mistakes culminating in dysfunctional protein and diseases. Improperly folded codons, competing codons, interference

41

RNA and micro RNA, just to name few of the possible road blocks in our endeavor, are presenting possible risk areas in any mammal genome sequence study. We hope to have given you a good idea on how this game of nucleic acids letters is playing around forming proteins for your body while you are oblivious of complex processes taking place in your own body. A couple of pages back we mentioned to you how this transportation system, seemingly so easy, can be threatened by simple small nucleotide polyphormism (SNP) and interference RNA, preventing genes from carrying out their job of protein synthesis. These S.N.P molecules love to mess around with the thymine nucleotide in the double helix and create unwanted problems for us.

Epigenetics in frontal pages

Epigenetic, the nucleotide process that may control a gene potential to express itself has come to take frontal page in some journals because it has been implicated in some common, but hard to break diseases. The most common and best studied form of epigenetic intervention in a DN A molecule is methylation. It has the potential to manipulate the cell to ignore any gene in a stretch of DNA. We are subject to epigenetic interference from very early in our life processes according to some researchers. It can begin during our embryonic phase of development. During this early stage of cell undifferentiation, let us say, blastocyst, stem cells accumulate, among others, methyl groups and direct stem cells into one of the three germs layers. Each of these germs layers produces a different set of adult tissues. These tissues we are referring to may be protein for your heart, lungs, eyes or brain. Making epigenetic a little more interesting, but headache provoking, a group of researchers at the University of Sydney in Australia discovered that methylation of the fur color (in mice) genes persist in the female germ line, allowing it to be passed down to off-spring like a change in DNA.[17] Local and worldwide studies share their common interest in the human genome, epigenetic, and particularly, methylation interference in DNA expression. A lesson to be seriously considered during disease research etiology is the role SNP in all forms play, including methylation, interference RNA, micro RNA during embryonic and fetus development. The cell has proven to be an excellent self-sustained unit of life, but there are many internal and external factors that threaten good health. Whether it is endogenous or exogenous, the great DNA molecule has survived millions of years. It developed a brain cells for intelligence and critical thinking to assist getting rid of temporal nuisance. This brain is developing very sophisticated tools for self-correction and to accelerate its own growth. Among the beauties of our recent technologies to improve our health is the decoding of the human and mouse genome a decade ago. Mouse and human have about the same organs, although they separated from each other some 70 million years ago. Most scientists agree that a mouse genome sequence is about 95 percent that of a human. We cannot play around with a human as a research scientific object, but a mouse can be carried with you even when you go on vacation. Gene-targeting technology using a strain mouse, we can knock-out a gene in any organ of the mouse and peacefully and intelligently, observe causal effects without any pain to a human being. Besides, a mouse does not demand dormitory, dining and recreational facilities. The researcher and his/her expensive tools is the expensive subject. These knock-out mice are very important in medical research. A mouse as an object for gene studies is relatively new. The first reporting of this technology stating that it could generate gene-targeted mice was published in 1989. Twenty one years later, three pioneer scientists working independently of each other were awarded the Noble Prize for Medicine or Physiology. The New York Times, Science Section, 10/09/2007, broke the news for us. The three

---

[17] - Scientific American, 8/20/2010, Vol. 303, #2, p. 2224.

pioneer scientists are: Martin Evans from Cardiff University, Wales; Mario Capecchi from the University of Utah and Oliver Smithies from the University of North Carolina.

Sequencing genome from different species can tell us how each one of us is different from each other genotype and phenotype. It can point out mutations and diseases. But, perhaps most importantly, comparing genome from different animals can tell us gene and gene segment that have remained unchanged for many years. Many years may be counted as meaning millions of years. We can date animal separation from each other based on a set of genes that have remained un-altered over millions of years. Observing genes location for protein expression placed side by side taken from different animals we may be able to tell how a species changed into a different phenotype. However, for us now, the immediate benefit of comparing genome sequences is the increased precision with which researchers can reveal the sequences that have been carefully preserved over time, implying that they have an important role in the organism. Alternatively, these comparisons can pinpoint sequences that differ in just one species or group of species.

Epigenetics

To some of you we may sound unnecessarily repetitious and perhaps, cognitively naïve. Our apologies to you, but our classroom experience and supervision taught us that going over a subject or a lesson is not a waste of time, but the strengthening of synapses. We will define or describe epigenetics as the study of heritable changes in gene expression that are not the by – product of changes in DNA sequences. We briefly addressed epigenetics indirectly when we mentioned SNP and gene modification. We also spoke about the chromatin complex composed of DNA and histones with its little tails extending out as possible SNP sites. We attempted to

44

connect it with current concepts of on and off protein switches. We could easily transfer this concept to our electrical system of on and off switches. It would sound and look neat, clean and easy to follow; however, we are dealing with living acids like DNA and RNA capable of forming multiple tissues and organs. During these processes, including before and after transcription, many modifications take place altering the ultimate by-product, protein synthesis. Where and how these modification take place has been the subject of curiosity and researchers during the last decade. Small nucleotide polyformisms molecules attach to chromatin, most often, linked to or hanging on to histones proteins, are called epigenetic marks by some researchers. Many unanswered questions are still lingering around the concept of the on and off enzymes switches. Are epigenetic modifications passed on down to off-spring through cell division? What about modifications that take place during the transcription phase where the two great acids, DNA and RNA are involved? Before we proceed to make more questions, which are not difficult to make, let us deal with chromatin modifications. One of the most frequently used concept to explain on and off switches is that a DNA and histones modification is consonant with either negative or positive transcription states. In histones post transcription modifications, cytoplasmic response to transcription factors- positive acting marks- are established during the gene activation. In histones post translational modifications is that in response to cytoplasmic signaling to transcription factors, positive acting marks are established... during gene activation by recruiting relevant enzymes by DNA bound activators and RNA polymerase. Similarly, negative acting marks are established across genes during repression by DNA bound repressor recruitment.

This seems to be a simple and attractive model to conduct chromatin research. However, some researchers claim that it is not as simple as it appears to be. On and off enzyme switches is not a matter of connecting and disconnecting cables or just turning on genes. Many of the marks we have been talking about play multiple roles. It is a dynamic metabolic process in intracellular chemical reactions that converts organism infinite number of molecules into energy producing action units. Chromatin marks are more complex molecules than anticipated; it responds to and performs multiple complex and different roles. It is now believe among some researchers that chromatin role in transcription is a mechanism in which modification establishes gene activation and then re-instate repression. Thus, DNA methylation and histone modification is a very dynamic process, it precludes a fixed status. Some marks now seem to recruit both, activating and repressing effectors proteins. Besides, positive and negative acting marks are established during transcription. Two types of chromatin modifications for the regulation of transcription of the protein-encoding genome are activation and repression. These DNA modifications seem to take place at specific junctions of nucleotides and at different locations.

Post translation modifications

There are a good number of post translation modifications. Most take place in the amino and carboxyl terminal histone tails. Among common modification sites, Lysine is a key substrate residue in histone biochemistry; it undergoes many exclusive modifications. This really seems to be the Achilles tendon we were talking about at the beginning of this essay. "In general, the functional consequence of histone post-translational modification can be direct, provoking structural changes to the chromatin. Changes can take place by recruiting proteins." For a complete picture on this subject see, Nature, 5/24/09.[18] Epigenetic changes are decisive, indispensable during the development and differentiation of different types of cells in any organism and normal cellular

---

[18] - Nature, 05/24/09, Vol. 447, #7143, p. 407-412, by S.L. Berger.

processes. However, like in all system, an inappropriate epigenetic modification can lead to serious health problems. Epigenetic processes may be involved in very sensitive areas as in a cell shape and function. It may intervene in chemical modification to the DNA and histones in the chromatin packaging. This macro-molecule-DNA- has taken appropriate measures to protect itself from harmful intruders, its job is awesome. The protein (histone) that DNA wraps around forming the chromatin structure, in itself has gone through multiple processes. Proteins are the most abundant molecules in our body. The number of proteins in our body makes it the most likely target for changes, modifications and assault by internal and external forces.

Epi=over or in addition to genetics, is now days used to mean the study of inherited- heritable- changes in our appearance. In genetics jargon, it is known as phenotype; the change may also appear in gene expression, meaning gene formation, but not involving changes in DNA sequence. At present time, the word heritable is paramount in epigenetic as long as it does not involve changes intrinsic, meaning, basic to DNA sequence. You may rightfully ask, how can it be heritable, if it does not involve DNA sequence?

Well, epigenetics has become a field of study of biology in itself. Epigenetics like RNA multiple forms and functions has attracted brilliant biologist researchers hoping to find and characterized toxic proteins that threatens our health. We have just finished sequencing the human genome; searching for tools to destroy pathogens that threaten our organism is our next challenge. The job ahead is worth our interest and enthusiasm. It took our most primitive cell millions of years to evolve from a simple cell to a complex organism like yours and mine. Among your body multiple organs, you have a brilliant brain that is engaged in finding solutions to diseases that threaten your life. Your brain, when not engaged in fighting diseases, amuses itself trying to become a painter like Picasso, a violinist or an astronaut navigating outer space.

Maternal nurturing and DNA methylation

Our main concern during this essay has been centered on how the genome sequencing can improve our health. There have been well documented studies in mammals reporting that maternal nurturing alters DNA methylation at the gene encoding glucocorticoid acid receptor. In the absence of appropriate nurturing, the chemistry of DNA in her offspring is seriously compromised or placed in jeopardy. If normal DNA methylation is negatively altered in genes coding for proteins destined for the hippocampus, the child's learning ability will be placed in jeopardy. The same will be true for negative DNA methylation for genes coding proteins for the amygdalas. In all probabilities, the child will not be alert enough to appropriately respond to environmental learning signals. Whether it is the hippocampus or amygdala deficient functioning secondary to an abnormal DNA methylation, the child's general health would be in great risk. Applying the definition of epigenetics as heritable changes, although not involving DNA sequence, helps us understand and deal with learning difficulties and emotional problems in our schools and general population. A considerable number of research studies have raised the profile of epigenetics making it a particular field of study in biology. Hopefully, it will help us understand and provide appropriate treatment for diseases like Parkinson, Alzheimer, schizophrenia, autism and violent behavior leading to bloodshed and death episodes that recently have taken innocent life in our society.

Cancer and epigenetics

A report from Epi Gen Western Research Group at the University of Western Ontario dated 01/31/ 2006 defined epigenetic as the study of heritable changes in gene expression that occur without a change in DNA sequence. By now, 2011, this definition has been established worldwide as the gold standard for research and findings on epigenetics. The leading author of this report, Dr. David Rodenhiser states that epigenetics mechanisms are critical components in the normal development and growth of cells. Epigenetic abnormalities have been found to be causative factors in cancer, genetic disorders and pediatric syndromes as well as contributing factors in autoimmune diseases and aging." This quotation places epigenetics abnormalities as a causative factor for cancer. Cancer has been under microscopic eye for quite a long time. However, there are different types of cancer appearing in different tissues and organs of the body. In man, we have prostate cancer that tends to stay localized, while in women, we have breast cancer that often metastasizes. We also have colon cancer that can be removed by surgery; we have throat and lung cancer widely associated with excess smoking. There is cancer whose etiology is attributed to environmental factors like toxic gases, air pollution, water contamination, pesticides, and even chronic stress has joined the list of cancer risk factors. However, most these cancers just named are of a temporal nature or etiology and do not classify in the definition of epigenetics because it is not passed on to offspring. The key word, heritable, cannot be applied to a cancer in a person with a long history of smoking a package of cigarettes a day. The same would be true of a person exposed to toxic or radioactive chemicals at his place of employment. There has been enough evidence to connect lung and throat cancer with cigarette smoking. The leading author in this article proceeds with DNA methylation and histone modification describing the basic components of a nucleosome, namely DNA wrapping around clusters (octomerous-from the Greek okto=8 and omers=parts) of proteins-histone.

This coiling of DNA around histones is widely known as the chromatin structure. Changes to this chromatin structure influences gene expression, the gene expression he is referring to is protein synthesis. You are aware that if protein synthesis is jeopardized, it would be a serious trouble for you and me. Protein shape determines function and, the shape of a protein whose final destination is the heart, specifically, a valve in the heart, most likely, if the fetus survives in the mothers' womb, as an adult, it will develop heart problems. We should add that cells have their own intrinsic repair system, but unfortunately, the repairing system does not work to our advantage every time we are in need. You can also take a protein destined to be part of an axon in a brain cell. Axons are engaged in electrogenesis; its functional components are sodium and potassium pumps that will send a chemical message across a synapse to a nearby post-synaptic cell membrane receptor. Any process or epigenetic mark that messes up the formation of proteins is threatening my life and health. Besides, you are already acquainted with the negative resulting effect of missing genes, segments of genes, the change of one nucleic acid for another and more. They all provoke health problems. Further elucidating genetic marks, the author names some of the enzymes involved in the process such as: DNA transferase, histone deacetylase, histone acetylases, histone methyltransferase, and the methyl binding domain protein, (the ending -ase denotes an enzyme). Dr. Rodehiser alerts us that alterations in these epigenetic patterns can deregulate patterns of genes expression, which results in profound and diverse clinical outcome. Some research finding in David Rodenhiser article called our attention because it has been confirmed by other research teams. DNA methylation involves the addition of a methyl group CH_3 to cytosine within CpG, it means cytosine/ guanine pairs. Cytosine is a nucleotide to both molecules, DNA and RNA. The point we are trying to show you is this epigenetic mark involvement in critical genes processes. Another emphasized point I like you to remember is that changes in DNA methylation may occur as a result of low dietary levels of folate, (a B vitamin), methionine, which is an essential amino acid, or selenium, which can have profound clinical consequences such as neural tube defects, cancer and atherosclerosis. This last observation was supported by five different researcher groups. We hope we do not sound excessively redundant, but among the highly suspected contributing factors altering DNA methylation and histone modifications contributing to the above mentioned health risk are air pollution, water contamination, pesticides, chronic stress, excessive exposure to dangerous chemicals and psychological causal factors. A note of reference to our epigenetic readings we want to alert you to in advance is that most of our quotations on epigenetic are dated up to 2006. A lot of research has been done and published during the last five years we hope to cover. Seventy years ago, who would have believed that at present time we can read the structure and instruction to put together a human being; make corrections on genetic diseases and create a synthetic bacterium? We are close to making science fiction a reality by creating artificial intelligence. Even splitting an atom was not only considered impossibility; but a threat to all forms of life on planet earth,

The DNA structure in the mind of our scientists

The idea of the DNA structure and function occupied the mind of a group of brilliant young scientist right after WWII including our star chemist, Linus Pauling. However, the Britons Rosalind Franklin, Maurice Wilkins, Francis Crick and the American J.D. Watson were determined to give the United Kingdom the prize of stardom in molecular biology. Although Rosalind seems to have been ahead of the group in designing the double helix for the DNA molecule, it was Watson who did not miss out a single opportunity to check out everyone else ideas and try to put it to work. He was checking out every article in biology and related sciences that might give him the final answer. He argued alone how the spiral strands should go, inside or outside, with its nucleic acids hanging on like necklace beads. On the other hand, Rosalind may have been closer than anybody else to the DNA double helix structure, but she insisted on more experiments before announcing it publicly. Watson disposition to share his curiosity and observation with Crick did not miss an opportunity to pick up cues from other researchers that might have advanced his own insights. Rosalind crystallographic pictures struck him like no other picture had. He finally figured it out that it must be a two strands structure with nucleic acids beads inside chemically attracting each other. His approving alter-ego, F. Crick, gave him his blessing and the structure of our genome soon was going to take front page on the most prestigious science journals, newspapers and television networks. I, for one, would have thought that Watson and Crick were celebrities to be paraded on Broadway in New York City and in Washington, D.C. Well, things do not always come to pass the way you like it to be. Even his Nobel Prize did not come until 1962. As if it helps anything, the person behind all these genetics, Gregor J. Mendel, a monk was not recognized even by his own colleagues in the monastery. In Watson case, researchers working on and around his discovery were awarded the Nobel Prize for subsequent discoveries relevant to DNA structure and function. Watson and Crick had to wait for another few more years to be called in from the Nobel awarding Committee.

What was so mysterious about this double helix structure? The solution of the mystery would unlock the key to the creation of life. They proposed that each nucleotide, namely, adenine, would chemically attract the opposite complementary strand nucleotide, thymine. Bingo, this is it. Now they needed to proceed with protein synthesis. How the four letters, ATCG of the double helix fit to arrange themselves to make the 20 amino acids necessary in the human body into proteins? Crick began to work with Sydney Brenner and together they demonstrated that the code was based on a series of three bases, named codons. Researcher Marshal Nirenberg and John Mathei identified the first letter in the genetic alphabet. They reported that the amino acid phenylalanine had the code, UUU. Please, turn to the table at the end of the book showing amino acids and their codons code. You can observe that most amino acids have more than one codon or triplets codes. In addition to the 61 codons for amino acids, there are three codons for a stop command. The stop codon indicates the end of an amino acid sequence coding for a given protein. Most proteins are comprised of 100 or more amino acids, and so a long sequence of codons is needed for any given protein."[19]

Watson and Crick parade in the Heroes Canyon

After discovering all the codons for each amino acids that will form peptide chains, which in turn, become proteins for cells and tissues in our body, what was next? In other words, it is a sequence of codons that codifies the production of a particular protein, which in reverse definition, it is a gene. These sequences of codon that codes for an enzyme will be regulating neurotransmitters in our brain. A neurotransmitter, like dopamine, which is involved in very serious brain diseases like Parkinson, Alzheimer, is also engaged in making us feel "good." This brain chemical is often associated with feeling "high" and addiction. This discovery opened up not only a new field of research- molecular biology-but another approach to therapeutic medicine. According to Crick Dogma: DNA-RNA-PROTEIN; it is the DNA nucleus that will be issuing orders for proteins synthesis in the cell. These proteins are the building blocks of our body. No wonder I was expecting a big parade

[19] - Brave New Brain, Nancy C. Andreasen, Oxford University Press, 2001, p.101.

for Watson and Crick through the Heroes Canyon on Broadway in New York City.

In retrospect, we thought that as soon as we had learned the step of codon sequence and protein synthesis, we could begin engineering or coax cell organelles and construct a miniature life system. I was not too far off; C. Venter in the year 2010 announced publicly that he had produced a synthetic bacterium. And...James Watson was around to celebrate it. Venter demonstrated that we have mastered the mechanism of DNA-RNA molecule secret to make live cells that will grow to become a whole beautiful organism like Dolly, the cloned sheep. Perhaps we will have to wait for Venter's disciples anywhere in the world to see the application of his newly discovered tool used for therapeutic medicine and pharmacology. We love being redundant; it is that in just a little over a half a century; we have gone from mule riding to supersonic air ships. When I was attending elementary school, a trip to the moon was thought of as riding on a broom stick. It was a witch story associated with evil spirits.

The very old RNA and DNA molecules

The DNA molecule began life many, many years ago either by chance or just basic chemical and functional need to pair with or incorporate another great molecule, RNA. That union facilitated, among other things, the formation of a protein and cell growth into a multiple cell organism; definitely, it was a giant step forward in evolution. Equally important was the annexation of another important molecule, the mitochondria. When and where these wonder molecules fused together, is basically an intellectual mental exercise. DNA survives by self-replication or duplication which is the same thing. During replication, it divides in two strands which are self-template for transferring the entire database into the new cells that will create new organisms. Second, DNA functions take place in the cell nucleus. A single DNA strand-messenger RNA- is formed which eventually will carry the message for protein synthesis at ribosome. Another RNA strand named transfer RNA will be traveling to a ribosome. Out of this assembly station, the ribosome will come out as peptide chains. This linear chain of peptides will be shaped into proteins for different organs and tissues in our body. There are many proteins performing many different functions in different parts of our body.

Introns and exons

Within the genome, we have coding genes and non-coding genes. The old name for non-coding genes was junk DNA. The new name for non-coding DNA is dubbed introns. In contrast, exons have the codes that determine the sequences of amino acids that will provide the final protein. This is not an easy job; just imagine the complexity of a template for a protein to my left foot toe and a template for a protein in the prefrontal cortex. As already mentioned on above pages, most of our genome consists of introns, or as you may say, junk DNA. Some researchers claim that over 90 percent of our genome is non-coding material. An interesting mechanism exists within each cell that during the transcription process that produces the mold for protein synthesis, the messenger RNA introns are thrown out, meaning they are removed before the protein is made. Introns do not have a job in the translation process. Researchers call the introns removal, splicing the introns out before they can mess up our protein assembly factory.

The transfer RNA strand carries three nucleotide sequence known as anticodon. It facilitates the matching of m-RNA and t-RNA at ribosome during the making of a protein. Just stop for a few second and think about all the different proteins in our body. Proteins that make up our brain have to even generate electricity to produce impulses that will secrete neurotransmitters that, in turn, will activate another cell.

Your memories from birth to death will be stored in proteins. And...how marvelous, the instructions or command from all these proteins, each one with a particular function, comes from the nucleus of the cell. This great molecule with its powerful nucleus is a very old one and has proven to be immortal. It has been feeding on chemical elements existing on planet earth for billions of years. Our body is a huge and complex chemical machine or factory that has survived earthquakes, tsunamis, hurricanes, electrical discharges of immense power, multiple iced ages and a huge meteorite about 69 million years ago. This cell is working from conception until death; it rest while we sleep although our vital organs are on 24/7 schedule, while genes are turned on and off as needed. Sleep is not an accurate term for us to deal with cells activity; take hibernation for instance, some small animals hibernate for months or even years, deep in the ground. This is an excellent research theme for sleep and insomnia studies. Somatic as well as brain cells are close to zero activation, but still alive. Buried deep in the ground they are not responding to outside stimuli like wind, rain, sunlight, drought, but maintained by a balance of inside earth temperature and whatever is going on above it. Cells are at a very minimum of activity to be able to survive when the appropriate time comes. Are they sleeping, unconscious, in a state of comma, hypnotized or just a bundle of atoms clustered in molecules that will begin chemical attraction with a little touch of nature?

Thalassemia and hematopoiesis

Genes may be turned on to produce protein to consolidate a memory permanently or to reconstruct a broken tissue-ligament- in my left toe. Genes may be called in to produce enzymes that will provide chemical reactions that in turn will trigger serotonin release and make me feel sad. All these switches –on and off- proteins may be provoked by internal or external causative factors. I have seen a puppy ran over by car trying to move while dragging its paralyzed legs behind him. I caught myself wiping out my eyes tears. My body was responding to a little dog in pain that I had never seen before. A visual stimulus had triggered a very complex response process I could not control at a conscious level. The cell nucleus was responding to, and commanding multiple actions that I was not aware of. Besides shedding tears for a little dog I had not seen before, I inherited thalassemia traits from my parents. I passed these traits on to my daughter. We were eleven children; five inherited it, one died before he reached three years old. In thalassemia, there are not enough red blood cells to carry oxygen. The protein hemoglobin that transports oxygen from the lungs to the tissues is made in the bone marrow by stem cells. When my body needs to make more of the protein hemoglobin, the pertinent or applicable part of the bone marrow DNA – meaning- the hemoglobin gene opens up, or as the professionals say, unzips. During this unzipping process, only one strand is copied; this process is also known as transcription. Here, a protein called RNA polymerase lends a hand in the process and, the single strand coming out of the bone marrow DNA is the single strand the original biologist called messenger RNA. This is the template for the production of a protein. We hope that in a near future my daughter and family members would benefit from gene therapy and eliminate that health risk from our protein factory. Protein is put together during the translation process. The command to produce protein ultimately comes from the nucleus, but what or who is responsible for alerting the bone marrow DNA that there is a need for more hemoglobin, this author does not know yet. We do know that proteins turn genes on and off at certain locations. We would like to quote from the New York Academy of Science 2005 report on hematopoietic stem cell research. "Despite the intense research, many longstanding questions of experimental hematology remains unsolved...Real time tracking of individual cells in culture, tissues or whole organism would be an extremely powerful approach to fully understand the developmental complexity of hematopoiesis. A vast array of cellular and molecular tools, RNA and protein expression data, and experimental methodologies have been developed for it and are already available for application. However, despite its therapeutic importance and longstanding research efforts, hematopoiesis research has been unable to satisfactorily answer many questions"[20].

Pasteur Institute in Paris

We have kept our interest on how genes are turn on and off meaning hoe genes are called to produce, in my case, hemoglobin proteins for oxygen transportation. We reasoned out that genes respond to internal and/ or external triggers or stimuli to produce a protein. In the process of protein synthesis, you must have an amino acid group at one end of the transference RNA. We looked for an amino acid in vitamins, minerals and dietary supplements rich in amino acids. However, our greatest concern was how to turn on the hemoglobin producing gene. At this level of research, we concluded that the concept of an on and off switch was very simple concept to rely upon. The whole process of protein synthesis involving many functional organelles within the cell points to complex system of cell metabolism and chemical intervention. While I was watching my son Joe work on an electrical switch in the basement of his home, I asked him how he would relate the electrical switch he was working on with a gene switch in our cells. He directed me to the Institut Pasteur in Paris. He added that while he was studying medicine in Spain, a friend of him, named, Alejandro, had gone to Paris for a short trip. His friend had good connections and was able to visit the Pasteur Institute. Working on a different type of switch was a couple of brilliant scientists, Francois Jacob and Jack Monod. Our intestinal parasite, bacterium Escherichia Coli, E

---

[20] -New York Academy of Science, Vol. 1044, 2005, p. 201--02

Coli for short, was receiving their undivided and focused attention. They found a repressor (a protein) that by binding to E Coli DNA, the repressor protein prevented the enzyme that makes up, or put together messenger RNA from doing its job. In guerrilla warfare, it means that this interference during normal protein processing, was a sabotage episode. The process of transcription had been had been interrupted by a molecule at a DNA switch. This was eye opening, if not brain boosting. Protein shape determines function. We concluded that a DNA child, a protein, could also play a role of regulating gene activity. This was another avenue of study, perhaps an RNA tool to engineer or manipulate our own genes. Monod and Jacob discovery made us even more curious, if not restless, about the behavior and function of this RNA molecule. It is playing many undisclosed games, right inside our body. These two scientists were working on my intestinal tenant, E Coli, and I was very happy to learn that soon, we may have good news on how to engineer it at will. How it lives in my intestines without paying rent besides feeding on my food. E Coli has become an excellent research model worldwide.

However, the ribonucleic acid was interfering with the building blocks of my body. For example, transference RNA can block protein synthesis, micro RNA can do likewise and much more. From the very beginning I became suspicious of RNA. Too much power in one person or object gives me goose pimples. Well, there was one person we could turn to, Francis Crick. Some researchers call him the father of the dogma, DNA-RNA- protein. Is it DNA that makes use of RNA to do his job, or is it the other way around? How would you answer this question? If you have any doubt, even peptide bond links --one amino acid linked to another-- is done by RNA. We rightfully respect DNA cell nucleus to send out messages and command orders for protein synthesis; but that there is an RNA molecule that can hijack the messenger strand and prevent the normal formation of a protein in my body; it was more than I could swallow, at the time. Anyhow, there are enzymes, again, produced on orders from the cell nucleus, that can influence gene expression. In fact, enzymes play significant roles in a cell metabolic and chemical reaction. In order to bring to closure the DNA- RNA dispute, we are going to refer you to a very reliable source that says, "The reality of the RNA World, an RNA dominated stage in the early evolution of life prior to the evolution of coded protein synthesis, has been firmly established by recent studies on the ribosome.[21]

Harry Noller and RNA

A scientist that could not resist the temptation of further studying the function of RNA, Harry Noller, stripped the protein assembly station, the ribosome, of all its protein. However, despite lacking the protein that is normally part of its structure, the ribosome was capable of forming peptide bonds. Returning to Jacob and Monod laboratory and their prized research model, the bacterium that made our intestine their permanent place of residence, E. Coli; they came to the conclusion that DNA must have a control system, actually, a protein to activate a gene, and tell it to get to work. Similarly, there must be another protein molecule to stop the gene from continue to produce more of the same. Thus, we attribute the on/off switch to these two gentlemen working with a parasite that made my intestines their home for life. We know that environmental factors can influence a protein to change its shape to provide a response to an external stimulus like very stressful situations. We also know that stress not only influences our responses through shape changes in protein, but most importantly, chronic stress can kill genes that produce the protein we need in our body. Our 45 years of experience working at a mental health clinic applying scientifically proven methodologies qualify us to write on harm done to a person

---

[21] -The RNA World, by R.F Gesteland, Cold Spring Harbor Lab. Press, 3rd edition, New York, p. 57

exposed to chronic stress. There are least three areas of the brain that are extremely susceptible to stress: the hippocampus, amygdala and the prefrontal cortex. These three organs or cluster of neurons are considered by most neuroscientist as part of the limbic system. As you already know, the limbic system is known as the home of emotions; it also happens that emotions are a very important attribute of human beings. Through emotions we socialize, makes friends, enjoy ourselves at parties, go on vacation to enjoy a couple of weeks away from stress at work, and through emotions, we fall in love, often ending in marriage. Another attribute that separated us from the rest of hominid is language. Through words we communicate our feelings, our emotions and in time, we were able to develop signs to write down our emotions. We keep our autobiography through episodic memories recorded in our brain, and as a family, tribe or nation, we pass on down our emotions in written form from our heroes. Emotions, if expressed in exaggeration, can be overwhelming and drive a person "nuts." When we fall in love it seems that the prefrontal cortex, our most rational cluster of neurons, is over ruled by the limbic system. Likewise, when anger takes over rational thinking, an individual may commit injury to himself or others.

You do not need us to tell you how a bugging stressor at home, job or neighbor can get your adrenaline level hijack your body and drive you nuts. The same argument can be applied to people predisposed to brain diseases like schizophrenia, depression and heart problems, hypertension, etc. Anxiety does not need a specific trigger to get you upset; we live in a world that is anxiety provoking 24/7. We turn the radio or television on, and stress triggers get you right on through your ears and eyes.

Thalassemia and Leukemia, my own trip

We began this journey on DNA and RNA not only because we are fans of Watson, Crick, Collins and Venter, but because they have helped us understand ourselves personally and our professional career. It has been our main concern to understand the person we see during our clinical practice as a whole entity. The person we see in our office is made of the same elements in my body. He or she is subject to the same daily threat and risk I do. We may ride the same subway; take the same bus to and from work, breath in the same polluted air, and often, eat the same fast food from the same restaurant. We live in cities with police cars sirens, ambulance and fire-fighters trucks alerting you to possible life hazards. Now, we come to a personal experience. We come from a family with a thalassemia minor history. Thalassemia is an inherited disease. There is no cure for this genetic disorder. The disease is classified as thalassemia major and thalassemia minor. I am a carrier of thalassemia minor traits. As a genetic disease I inherited from my parents. What concerns me most is that about 50 percent of my family, 5 out of eleven children were positive for thalassemia minor. Jack, my brother just immediately before me died before he reached three years old. According to my mother, he died of anemia. My oldest brother, besides being positive for thalassemia minor, died of leukemia after he reached sixty years old. Two of my oldest sisters had children and grandchildren that inherited thalassemia. My oldest sister grand daughter died of leukemia at age 9. My second oldest sister had her grandchildren followed up at a local health clinic for thalassemia minor. Both, thalassemia and leukemia are blood diseases. Therefore, the suspicious that there may be a connection between this two fatal blood disorders haunted me for some time. In thalassemia we know that it is a dysfunctional or a mutated gen. However, in leukemia, a blood cancer, like all cancers, the causal factor or factors for this fatal disease is not fully understood. Some studies involved some genes in it; however, there are many different type of cancer that makes it difficult to point out its etiology to any single gene or group of genes. As we have said earlier, cancerous cells are rebellious cells that proliferate and invade and destroy healthy tissues and organs in our body. The best tools in our modern medical technologies have not been able to identify and explain what originated the initial proliferation of malignant tumors cells. There are many endogenous and exogenous trigger factors that it has been very difficult to point a specific or single causal factor. Does a breast cancer have the same etiology as leukemia, colon or prostate cancer? People exposed to pesticide, toxic chemical or chronic stress can develop cancer; can you name people that are victim to the same etiology?

The Bubble Boy, David

In our case, it was a blood disorder; in thalassemia we do not know if it is a mutation or deletion of genes. We know that we do not produce enough globins to carry this vital element, oxygen, to all parts of our body. Otherwise, our family has been healthy; my father and my mother lived to be almost a hundred years old at their time of death. There is a movie that pops up in my brain, the Bubble Boy, a boy named David that was born without an immune system; his body was unable to make antibodies to destroy intruders to his organism. David lived in a sterilized environment (the bubble) until he was 12 years old. A bone marrow transplant from his sister did not help. A virus was present in the cells she donated and killed David. He did not have T & B cells to alert and organize a host of other auxiliary and killer cells that would protect him from a virus.

The literature that we researched claims no linkage between thalassemia and leukemia. For thalassemia disease, gene therapy may come with a possible solution. Trials with human have not produced the desired response, but we were not born flying. Whether, stem cells including induced pluripotent stem cells, embryonic cells, gene therapy, micro RNA or cloning, definitely, we are going in the right direction. We will convert blood stem cells into T cells progenitors and join B cells in the battle against pathogens. Our initial concern over the etiology of leukemia was partially answered during an international symposium in Germany in 2004. It reads as follow: "Several studies reported the derivation of leukemia stem cells from hematopoietic stem cells as a result of epigenetic events...[22] Whether in the U.S.A., Europe or eastern countries like Japan, South Korea or China, embryonic stem cells research as well as induced pluripotent stem cells will be in the forefront of medical research and pharmacotherapy. Under IPS we can create any cell in our body. It is even possible to build tissue and organs for a body in need of repair. The most beautiful part of IPS technology is that the problem of rejection by our immune system will be overcome because we will be using stem cells from the same person in need of repair. Our organism is a factory of functional cells. Japanese and American are at present time the leading researchers in the field. We will not be doing anything out this world, we have machinery inside us. Watson and Crick placed us on the road to decipher the human genome code; now we can enhance our brain to absorb and intelligently use all this technology to improve and accelerate brain cells that have moved us ahead of any other hominid.

Learning from Watson genome

What can we learn from Watson genome sequence and the millions of small nucleotide polyphormisms that are attached to the three billion pairs of the genome? James D. Watson genome sequence showed about 3, 3000,000 SNPs or simple substitutes of one base for another at a particular site in the genome. Watson genome variation as well as that of Craig Venter seems to be typical of a genome variation of individuals of predominantly European ancestry. Of the 3, 3 million SNPş, 10, 654 cause amino acids substitution within the coding sequence. In addition, the leading author of this article, David A Wheeler, wrote that his team identified... copy number

---

[22] -N.Y. Academy of Science, op cit. p.1X

variations resulting in the large-scale gain and loss of chromosomal segments ranging from 26.000 to 1.5 million base pairs. For an untrained individual alien to genome sequence, these are a lot of bases substitution and gain and loss of base pairs. Uneducated and un- accustomed to see so many changes within a single genome; it might be a little scary. However, it happened to a Nobel Prize winner and brilliant scientist, J. D. Watson. The point we try to convey is that so many variants did not affect him mentally or emotionally; he is mentally, a well-rounded individual. We have the genome sequence of two more brilliant scientists, C. Venter and Francis Collins. This author has not read enough about the genome of these scientists, but would not be surprised to see similar patterns on it. M.V. Olson commenting on Watson genome wrote: "The genome sequence seems to show that he is a carrier for a handful of mutations…but these mutations have no known effects on him." Watson, along with F. Crick, gave us the double helix structure over half a century ago. Now, these brilliant and remarkable men leave us their genome sequence for us to learn about ourselves through millions of gene alterations found in their genome. We close Olson teaching lesson with another quotation from him: "The challenge in human genetics now is to learn how to correlate genotype with phenotype (the external appearance) with special attention to disease predisposition and response to therapy."[23]

DNA, SNPs and disease

A brief comment on the above is not out of order in here. Companies that advertised to get your genetic code decoded and reveal to you how you got your beautiful legs from or how to change the color of your eyes is more of a publicity stunt aimed at profit making than scientific accomplishment. However, there are sophisticated and reliable companies that perform genome scanning that reveal to you

---

[23]- Nature, Vol. 452, 04/17/2008, p. 819-20 and 872-876.

potential assets and liabilities in your body genome and you may plan accordingly. Again, a word of caution, interpreting for you the potential assets and liabilities that may come in millions is another big question. Among those genomic markers are SNPs (short nucleotide polyphormisms). They are DNA variants that in some cases are associated with your susceptibility to a disease. But, of course, because these SNPs show up in your DNA, perhaps in the wrong place, it does not mean that you will get the disease. We like to remind you once more that both, J.D. Watson and C. Venter sequenced genome show liabilities; however, generally, both are healthy and brilliant scientists. Both have been a blessing to humanity. Take for instance, variants of a gene dubbed APOE associated with Alzheimer risk. Are SNPs like structures, also variants that contribute and determine the degeneration on neurons that ultimately end in Alzheimer? How are introns, the non-coding genes, and their variants related to the disease? How does exercise, meditation and a mentally busy and well-focused brain like those of Venter and Watson influences the growth of these variants as a liability or assets? How these companies interpret the propensity to a specific disease with so many epigenetic and environmental risk causal factors jumping on us 24/7? Besides, a good number of these findings are based on statistical average and there are a significant number of samples that do not fit or fall within the average category. We do not know, but if we were to go by the average sample, let us say in scientific talent, John Nash, C. Venter, J. D. Watson, A. Einstein and even Da Vinci four centuries ago had to be counted out. I am not throwing out the average scale. What I am saying is that before you lose sleep and money over your DNA variants; please pay a visit to your physician or a qualified health professional.

Genome surprises

Genome sequencing has brought us many surprises. We have sequenced a few animals including humans from European ancestry, Africa and China that have provided us with insight into our own evolution and development. Among our surprises was the incredibly large amount of non-coding genes taking space inside our skull. Besides SNPs, epigenetics, mutations, misspelling and others, we have learned that a mutation- a change of a single letter of DNA code- is not that bad at all. During gene duplication some will be ill prepared to function normally and will be degenerated. The cell has its own self-correcting mechanism. However, an occasionally produced mutation may slip by and allow a duplicated gene to prove itself by doing new and useful job for the good of the cell, and ultimately, your organism. A gene may strive to best serve the cell and may give rise to competition of functions. Smell and sight in dogs and birds might well illustrate our point. We cannot forget that shape determine function. Many things take place during cell specification that we cannot elaborate in this essay. Here, we are not talking about the battle of the fittest, but which shape fits in and serve best the cell and organism. "We humans have about 400 genes for smell receptors alone, all of which derive from just two in a fish that lived around 450 million years ago.[24] This is not a short trip by any means, but there is not a way we can compare our sight and smell with that of an eagle, a hawk and dog.

During the last decade we have been looking deep and close into the intricacy of the genetic code of life. Letters change within a gene, misplaced codons and introns into promoters domains, as well as DNA and histones modifications can interfere during protein synthesis, embryonic and fetus development, and consequently, provoked malformations and

---

[24] -New Scientist, 11/22-28/ 2008, p. 44.

diseases on us. Even a complete chromosome can be duplicated as it happens in Down's syndrome. In 2008, biologist Wen Wang in China working with several species of fruit fly, an excellent research tool, was able to identify new genes that have evolved over millions of years. To his surprise, he found that 10 percent of the new genes had arisen through a process called retroposition. This occurs when a MESSENGER RNA copies of genes- this is the RNA template sent to the protein assembly factory, the ribosome; these copies of genes are turned back into DNA and inserted somewhere else in the genome. The interesting thing for us in this discovery is that the gene copies created by retroposition are not the same as the original ones. We know that genes consist of more than just the sequence coding for a protein. Among laborers involved in gene and protein synthesis are promoters that will send signals for additional participating molecules.[25] Promoters, as gene regulators, are something we have to reconcile with; they are involved in determining the location, time and volume of the protein to be made. Promoters are facilitators that we need to fully gain to our side and convert it to first class research and therapeutic tool. My friend, the inquisitor, just whispered into my left ear: What you need is a tool to eliminate most junk DNA, place your limbic system under strict control of your prefrontal cortex and build a synthetic pathway between the hippocampus and long- term memory storage you can turn on at will. I said to him, How about T and B cells in our blood system and glia cells in the brain? Whether we can insert a synthetic gene in a chromosome, as far as we know, is science fiction at present time; but gene duplication seems to be as common as mutation which is changing a single letter of DNA code. During gene duplication and reshuffling of its component parts, the resulting "new gene" may well be a different gene from my own ancestor pool of genes. Genes are the instructions to make proteins; and proteins are the basic pieces of matter nature made it necessary to build your body.

---

[25] -New Scientist, 11/22-28/ 2008, p. 46.

Furthermore, genes hold the secret of your entire life ever since you existed as a simple organism. By studying your genes, you will learn about all the secrets your own ancestor did not share with you.

The last ten years in genome sequencing

A lay person overview of genome sequencing during the last decade ending in February 2011 would include speculation about protein coding genes from over 150, 000 to about 22,000, junk DNA, epigenetics, and probably, its similarity with other mammals like the mouse and a chimpanzee. However, Professor E.T. Dermitzakis, from the Dept. of Genetic Medicine at Geneva, Switzerland points out that in the genome, the exact location of regulatory regions was unknown, and only a small fraction of the variations existing within the human population had been characterized. The basic components in each genome are largely the same, but the way they are used differs from tissue to tissue and from person to person. Understanding the rules of gene regulation, the grammar of the genome, is key to the understanding of the human body.[26] Professor Pardis Sabeti from Harvard University said: During the pregenomic era, evolutionary genetic was a painstaking process. Scientists hypothesized instances of selection and sought confirmation case by case. As of the year 2,000, only a handful of cases had been identified. Technological and analytical advances in the past decade have enabled us to progress from hypothesis testing to hypothesis-generating science…We can scan the entire genome to identify variants under natural selection. [27]

Chief Scientific Officer, Eric Steps from Pacific Biosciences of California said, "The first human genome sequence, published in 2001, provided a canonical reference from which to understand genome structure, as well as registry

---

[26] -Science, Vol. 331, 02/11/2011, p. 689.
[27] -Science, Vol. 331, 02/11/ 2011, p.690.

of functional units...At present, thousands of genes that influence susceptibility to hundreds of diseases- associated phenotypes has been identified. [28] Science fiction is fiction until we learn to do it. I hope I can quiet down my inquisitor for a while.

Watson, Crick and Venter chemistry served them well.

In general parlance, J.D. Watson genome sequence showed us that his biochemistry served him well. Proteins, amino acids, codons, genes, DNA and RNA that we have been talking about, he left it for us in his genome. It comes from a person dedicated to work for all of us. It served him well producing a brilliant scientist that locked his chemistry with another jovial and equally smart person, F. Crick whose chemistry connected very well with that of Watson. Neither of the original players of the double helix structure stuck together as Watson and Crick did. We wonder to what extent Watson's SNPs and mutations, all parts of the dynamic of our chemistry, played a role in his behavior and stamina to stay on course and ignore road blocks encounter during his challenging life trip. We remember the biography of another outstanding scientist of the XIX century, Madame Marie Curie, that despite prejudice and male chauvinism, she remained loyal to her ideal and love for science. Her "chemistry" will endure for centuries to come. She died doing research in her efforts for us to understand ourselves and live a happy life. Her laboratory was her universe, and everything in the universe was subject to scrutiny and analyses. Cause and effect was the golden rule to do research. Today, we do research with models that can be observed and replicate in order to make use of our laboratory findings for the benefit of the general public. One of our most widely used research model in medical research is a mouse. Its gene is very similar to human genes. Most recently, 2011, an orangutans genome was sequenced and found that its genome

---

[28] -Science, Vol. 331, 02/11/ 2011, p.691.

is 97 percent that of humans. We already know that our next closest relative among hominids, the chimpanzee is 98.4 percent like us, humans. You may wonder how the sequencing of these animals helps us medically and otherwise. Just think about chromatin methylation and modification and other relevant epigenetic causal factors in human maladies that may be clarified through the sequence of animals with a genome very close to ours. For instance, when we silence a critical gene in a chimpanzee we learn how it affects human health and behavior. Consequentially, we can develop medical therapeutic tools to either correct the gene and its proteins or arrest he disease.

Gene targeting is not Sy Fy

Gene targeting does not belongs to science fiction television shows to get people adrenaline run high in their bodies. Our preferred animal of research model the mouse, has the same organs as humans do, and their genes are about 95 percent identical as any male or female specimen of our human race. Using gene targeting technology, scientists the world over, can knockout, meaning, turn off, suppress or totally inactivate a gene in a mouse mimicking a human diseases and apply therapeutic technology that may be used on real human diseases. Scientists can force a mutation gene to reappear at specific organs and tissues in our body, including our brain, and observe its behavior as it relates to a particular illness. The knockout technique has provided genome researchers, and above all, medical research, the opportunity to identify and observe each gene in our body through the mouse model from birth to death. It is estimated that we have about 24,000 genes. You may argue that with our modern technology, learning everything about our genes should not be an insurmountable challenge for us to identify disease risk points in our genome. This is exactly what we thought when we first started researching genes. What we had not anticipated was the role of junk DNA, DNA methylation, histone modifications, and in general, all epigenetic and environmental factors contributing to the etiology of any disease in our body. There are millions of tiny nuisances in our human genome that seems to be uninvited guest that have made our bodily organs and tissues their permanent home. But thanks to our brain, we are determined to get to the mountain top and get control of our destiny and health.

The science king of the XX1 century

Nanotechnology, our most recent brain child, has engineer tools to look into, and possibly repair not only dysfunctional genes, but go into molecular and atomic levels. It is the science king of the 21 st centuries, and unless we use it against each other, will change the way we see ourselves and the world around us. In a quarter of a century of research, we went from a computer that needed a truck to move it around, to something as small as your wrist watch. We have discovered materials stronger than steel, but lighter than a feather. It is electronic propelled by tiny vehicles capable of achieving light velocity. Sun energy will be used to energize traveling miniature vehicle carrying human cells with memory carrying proteins. Cells will replicate at sites chosen by humans on planet earth. Hibernation will be used as an adjunct technology during the voyage. This technology is not in Sy Fy books for recreational purpose, it is in our laboratories. The suspected anticipated problems do not lie in our brains; it lies within our economic, social and political systems around the world. How to feed and control around 10-11 billion people on earth within next forty to fifty years is our greatest challenge. A recent article in a prestigious scientific journal wrote that feeding 15 billion people by the middle of the century would be "easily possible." Easily possible if it brought more land into use…but that land is in Africa and Latin America. A problem already existing in those places is not mentioned. Land in Latin

America and Africa that in the past was used to feed people, is now being used to produce some sort of fuel for the developed world upper class ride their luxurious car, yachts and airplanes.

Communication among people would be much easier and accessible to the majority of the population in any given country. Migration around the world will be an enormous problem most difficult to control. In addition, the world areas for future agricultural use, is exactly the place for very fast population growth. India and China and countries between those two, have about a third of the world population. They will be moving out of their geographical boundaries rather soon, and there will not be a power strong enough to challenge it. They are already doing it in many parts of the world. Western Europe seems to be inundated with migrants from Eastern and African countries. They arrived by homemade boats, walking, and trucks, and as "tourists". There are justifiable arguments that we will come to some sort of agreement for self-control of unrestrained birth. It is a little over a half century ever since we established the United Nations to avoid wars. You pass judgment of its mission. The Korean War, Yugoslavia, Iraq, Afghanistan, Israel- Palestinians and Lebanon have been fighting each other ever since the U.N. was established. Now, imagine how we can get together to solve a minor problem when the issue is food on the table. Agreeing on a single subject has not been a strong point among humans behavior. Man spends more money on defense weapons (which actually is nothing more than a weapon to kill the neighbor) than food, health and housing.

It is not a brain problem

The U.S. Department of Energy has launched a late technology initiative to convert sunlight into hydrogen and other fuels. Its director, Nate Lewis said, "We have to scale up from nanoscale to macroscale." Hopefully, it will capture photons and produce electricity, which can be used to split water molecules and produce hydrogen. With artificial photosynthesis, photons from the Sun would drive a wireless chemical conversion process to generate fuel.[29]

Converting land dedicated from millennium to produce rice, corn, wheat, bananas, etc., into fuel for the super –rich is not the solution to the above mentioned problem. The answer already exists, and we thank our brain for it. Genetic engineering seeds, animals, fish and plants have been among us for millennium. But, at present time we can do it faster and more abundant. We engineer seeds to fight insects, and survive during drought. Instead one crop a year, we make it for two crops a year. We also improve the size of fruits, vegetables and roots. Most of it has come out after we introduced the double helix and began sequencing genomes from animals and plants. Genetically engineer maize and Soya bean has been one of our greatest feast of food engineering. Conversion of this land to provide fuel to move vehicles for the rich is not going to solve our world problems. The sun is our greatest source of energy we need to explore, besides, how about hydrogen from our oceans? We will have to invest on research with little initial profit, but the final solution will save us all.

DNA-RNA- Amino acids –Proteins Research

DNA, RNA, amino acids chains and proteins research has kept our mind focused for many years. The double helix comes even in food boxes. How proteins are made was a curiosity we could not put away or silence not even during sleep. Researchers have helped us during this dream trip by discovering the basic RNA molecules involved in protein synthesis. We learned about r RNA, t RNA, m RNA all involved in the making of a protein. Interference RNA is another molecule that becomes a nuisance when it interferes with forming a protein molecule. We all know that proteins are

---

[29] - Nature, Vol. 466, # 7306, 07/29/10, p. 541.

peptide chains with the above ribonucleic acids molecule mentioned above playing a significant role in carrying or transporting its components polymers from the cell nucleus to the final assembly station, the ribosomes. Recently, Phillip Kapranov et al. in the journal, Nature, 7/29/10, suggested a novel RNA copying mechanism. He claims that RNAs are not merely degradation products of longer transcript, but could indeed have a function. Profiling of RNA, including the sRNAs, can reveal not only novel transcript, but make clear prediction about the existence and properties of novel biochemical pathways operating in the cell. This article appealed to us because, among other things, we have maintained that RNA is our body mechanic that put our parts together, although, after receiving the message or order from the cell nucleus. RNA folds itself in many shapes carrying multiple functions that make us show more respect and admiration for this acid than ever before. Some readers may think that we are overly concern about i RNA as a "spoiler" in protein synthesis. Their concern might well be well justified, but look at the pharmaceutical and investor corporation that wants to put i RNA into a therapeutic tool to take care some of our illness. RNA i was discovered in 1998 and in 2006 the scientists that discovered it were awarded the Nobel Prize. Soon thereafter, Merck, the giant pharmaceutical company paid more than one billion dollars in its competition to make drugs out of i RNA. Another giant in the drug business, Roche in Switzerland gave up in this competition after spending more than ½ billion U.S.A dollars on the technology. Furthermore, listen to Heidi Ledford writing for Nature, The development of RNA i based drugs has stalled as companies confront the challenge of delivering RNA molecules, which are notoriously fragile to target cells in the human body, and then coaxing those cells to take up the RNA. "Getting these molecules exactly where we want to go is a little more difficult than originally thought." added Michael French, chief executive of a Biotech company in Bothell, Washington.

Equally curious and interesting for us is our body building block, the protein. There are some proteins and their receptors that get a lion share of our attention. Naively, our attention has been geared at our brain cells and membrane receptors. Heidi Ledfor writes: about one quarter of approved drugs target members of a single protein family, the G protein coupled receptors. She further added: Members of this clan- the largest protein family in the human genome, control everything from hormone signaling to the perception of light and scent. A no small amount of money from the U.S.A., $ 290 million will be re-directed from learning amino acids folds that comprise a protein... to solving some of the world most troublesome and medically relevant proteins including the G protein coupled receptors. Besides this protein, mitochondria as well as protein that regulate gene expression will be principal targets. We feel relieved by this announcement. The focus will be on proteins, and specifically, proteins that are medically relevant. We want to know not only how proteins are formed, but how they relate to and become the causal factor of many of our health problems. You already know that protein packaging gives it shape, function, as well as the length of the chain of peptides, in addition to multiple containing subunits, provide for its different roles in cells, organs and tissues of our body.

A late word on protein synthesis before we pass on to the next paragraph is necessary. The mantra or dogma among biologists is that a protein needs a structure to be able to function properly. Every book I read emphasizes that function demands structure. There have been serious arguments that even the most unstructured ambient is ruled by some type of structure and order. Our personal opinion is in consonant with structure, function and order. Peter Wright, a protein biophysicist at the Scripps Research Institute in La Jolla, California told a group of scientists attending a meeting of the American Association for the Advancement of Science in Washington, D.C., "The recognition of disorder has grown dramatically." Data are fast accumulating from all fronts-biophysics, bioinformatics and cell biology in support of widespread of disorder. P. Wright and team member J. Dyson back in 1999 wrote a review pointing to the growing collection of protein that seemed to function despite their disordered state. However, there are some researchers with a broad imagination like Joel Janin from Gif-sur-Yvette in France who says that "the whole concept of disorder seems incompatible with the lock and key model. You might as well try to open the door with cooked spaghetti." Little by little, a fundamentally new structure of relationships between protein sequences, structure and function is emerging: a continuum running from the most rigid lock and key enzymes and molecular machines at one extreme through to durably unstructured spaghetti, spanning all degrees of structural ambiguity in between."[30] How this cooked spaghetti structure will work or is already functioning, is an enigma for us, but will keep an open mind and see how it turns out. When Rosalind Franklin, J. D. Watson and F. Crick were attempting to put the double helix together, doubts seemed to outweigh good intentions.

G proteins receptors

What are G proteins-coupled receptors? It is a large group of proteins that are attached in the plasma membrane of most cells. They seem to look for, and attach to agonist molecules responsible for conveying sensory information. This receptor is also involved in the transmission of changes in physiological states of the cell. Once the receptor is activated, G protein initiates signal cascades that control multiple biological processes. How best approach this challenge is going to require elaborate team work. The mouse and the petri- dish are excellent research tools, but translating its finding to a protein receptor physiology is another issue. The cell internal environment is aqueous complicated by internal chemical

---

[30] - Nature, Vol. 471,) 3/10/2011, p. 151-53.

reactions; likewise, the outside is equally influenced by an aqueous environment. Protein shape and function are conditioned, among other things, by its aqueous environment. How water molecules come in contact with protein in different cells, affects proteins functions and related medical problems. The dynamic exchange of water and protein atoms and molecules exchange in humans at different stages of development is another challenging issue. This huge family of protein receptors is responsible for many functions we are not equipped to fully elaborate on. However, missing a chance to research the most recent and relevant literature available is something we could not avoid. This is not a local issue, but a worldwide challenge.

A G protein coupled receptor is essential for Schwan cells initiate myelination. Signals from the axon activate expression of specific transcription factors such as oct6 and kro20. I write this names and number just for reference purpose, not to bug your head. These two molecules begin myelination in Schwan cells. These are glia cells that produce myelin in the peripheral nervous system. Hold on, I will explain it for you. Myelin is a lipid or grease or fat that is wrapped around the axon; it insulates the axon and allows the electrical message (yes electrical message) reach its target fast and uninterrupted. Lack of myelin in the neuronal axon can be fatal. It may provoke serious illness such as multiple sclerosis, amyotrophic lateral sclerosis and Krabbe. A type of brain cell called glia oligondendrocytes produces myelin. In your brain you have cells known as neurons. Neurons are composed of a body which is round disk like with many processes or fibers extending out. These extensions are called dendrites and receive messages from other brain cells, in particular, neurons. There is another process or fiber coming out the cell body called axon. Messages from the neuron nucleus go through the hillock into axon. A hillock looks like a ring linking the cell body and the axon. The axon is wrapped around with myelin almost like plastic is wrapped around an electric cord or wire. There is a difference between the axon wrapping and the plastic covering an electrical cord. The axon is not totally covered by myelin; there are small segments or spaces in the axon that are naked or free of myelin. The naked spaces allow electricity in the axon to gain speed jumping from one node to another down the end of the axon. This electrical message or impulse will provoke a chemical reaction, meaning, a neurotransmitter will be released at a synapse. A synapse is a space between each neuron in the brain. Among the neurotransmitters released into the synapse are serotonin and dopamine that we talked about. Loss or destruction of myelin most common in the central nervous system is in the brain, spinal cord and optic nerves. Multiple sclerosis is the primary disease of demyelation. Some of the diseases caused by destruction of myelin that we know of are: multiple sclerosis, Krabbe Tay- Sachs, phenyl-ketonuria, Hurlers syndrome and neuro-myelitis optica. Of the above diseases related to myelin loss, I have had professional experience with multiple sclerosis and Krabbe.   M.S is characterized by recurrent episodes of demyelation. It is a chronic autoimmune inflammatory and neuro-degenerative disorder of the central nervous system. The duration of each attack and how often it takes place do not seem to be well established as yet.

Dies Meijer writing for the journal, Science says: "Myelin is laid down and maintained by dedicated neuroglia cells- oligondendrocytes- in the central nervous system and Schwan cell in the peripheral nervous system."[31] My experience with Krabbe was a very sad one. A little girl 18 month old otherwise considered healthy by the local pediatrician began crying without any obvious provocation. Her mother took her to the local physician, but to no avail. She was later taken to a specialist who diagnosed her as Krabbe disease. She was later transferred to a well-known pediatric

---

[31] -Science, Vol. 325, # 5946, 09/11/2009, p. 1353.

hospital in the East Coast of U. S.A. Her initial symptoms were: feet turned inward, unable to stand up, could not walk, and complaining of pain all day long. Soon afterward, the little girl began to lose sight and hearing. After a few weeks in the hospital, she was discharged home to her mother care. She is in a wheel chair with weekly nursing visits and monthly physician checkups. Another reliable resource on myelin and Schwan cells writes: " The axon of many vertebrate neurons are enclosed by a chain of supporting cells called Schwan cells in the peripheral nervous system, that form an insulating layer called the myelin sheath."[32]

A personal encounter with histamine

During my clinical practice I have seen many people with rare symptoms and diseases that have strengthened my dedication to assist individuals in pain. One area that always challenged my ability to help anyone is allergies. We have been fighting allergies for a long time. From dust, pollen, weeds, cats, dogs, food and, even an occasional rare perfume would provoke a wheezing sound, running nose and watery eyes. A visit to a physician usually brings relief after taking the prescribed medication. However, the prescribed medication does not always work. We should not blame the medication or the physician. Although we are made of, basically, of 20 amino acids that will be formed into proteins, the end product of amino acids and proteins, your body and mine, are not truthfully equal. We end up uniquely two different individuals responding to internal and external stimuli in different fashion. However, despite our differences in behavior, there is always an average response. And…a medication or rather the usefulness of any medication is measured on the average individual, not on the difference between each individual person. Consequently, a medication that worked wonders in

---

[32] -Biology, 3rd Ed., Neil A. Campbell, the Benjamin/ Cummings Co. N.Y., p. 984.

Peter not necessarily will make the same miracle for Joe or John. Remember, we carry our genes from our ancestors, but we are also greatly influenced by epigenetics and our immediate environment. My mother and two of my siblings were always complaining about allergies. Allergies have been my inseparable companions for many years besides having a relatively poor digesting system. Greasy foods like pork chops, milk, cheese and even butter pecan ice cream that makes my mouth a lake of saliva, do not fall alright inside my stomach. Somehow, those protein-enzymes responsible for breaking down fat are not helping me. I avoid those foods although my eyes and internal voice called me a coward for ignoring my desire to enjoy myself. From the health food store I buy appropriate enzymes. Besides allergies, I have learned about another health nuisance, histamine. We have said that our organism is a very complex chemical reaction factory producing millions of reactions that we are committed to fully understand. You are correct when you say that many of those chemical reactions are the effectors of some of our diseases. When you experience an excess of gastric fluid in our stomach and feel that bitter taste, you know what it is that I am talking about. Health foods stores and over the counter medication at pharmacies, have full shelves waiting for you to choose from many similar remedies. To calm down my stomach acid and running nose, I carry with me antihistamine called, Diphenhydramine.

Histamine as a chemical messenger

Histamine is a chemical messenger that is responsible for handling a wide range of cellular responses. We have been talking about DNA and RNA molecules going through multiple reactions until proteins come out to form organs and tissues. Take a minute and visualize all the chemical reactions going on in this precise moment inside your body, and specifically, inside cells. Among the reactions we are referring to are allergic and inflammatory reactions. Most likely, you must have had an allergic reaction of some sort. What is not visible and difficult to explain is the inflammatory reaction taking place inside you. Some asthma attacks could be an allergic reaction of the lungs. Besides or along the inflammation on your lungs you may have an excess of mucus that makes it very difficult for you to breath. You feel like choking on your own saliva and mucus. You cough and cough again and again to clear your throat and breathe freely. In addition, and to my surprise, there is an excess of gastric fluid in your stomach threatening you too. You try to cough out all the mucus coming out your lungs, but from inside your stomach comes out a full cup of vomit, as bitter as nothing you have tasted before. All this pain and discomfort was basically provoked by histamine or an excess of it. This chemical messenger mediates a lot of responses in millions of cells in your body. It is suspected that it works in brain cells too. Besides all the problems that an excess of histamine can create for you, it is a powerful vasodilator. We do not chicken out because some illnesses are hard to crack down. Take for instance, serotonin; it is highly involved in brain functions. It is a neurotransmitter involved in depression. However, the largest amount of serotonin is found in cells in the intestinal mucosa. "Serotonin has multiple physiologic roles, including pain perception, normal and abnormal behavior, including affective disorders, and regulation of sleep, temperature, and blood pressure."[33] Besides our well known serotonin, we have another public friend that most people talk about even over the dinner table, dopamine. This neurotransmitter is a chemical that make you feel extremely happy when produced in excess. It is also involved in two chronic and degenerative brain diseases, Parkinson and schizophrenia. In Parkinson disease, cells producing dopamine in the brain substantia nigra died out. Recent medical technology is helping people with Parkinson. In schizophrenia,

---

[33] -Biochemistry, P. C. Champe & R. A. Harvey, 2nd ed., J.B. Lippincott, Phila. 1994, p. 265.

there are some genes involved in it, but this disease is more of a syndrome than a single encompassing behavior. There is good medication to control most of the symptoms of schizophrenia at present time. Schizophrenic symptoms are not difficult to identify by health professionals. Youngster from ages 16 to 21 are very vulnerable to strong stressors and those individuals with a schizophrenic pre-disposition need extra help from parents and teachers.

Eric Kandel on schizophrenic symptoms

Dr. Kandel and Eleanor Simpson, both neuroscientists at Columbia University in New York City reported in 2006 discovery made on D2 receptors associated with schizophrenia. Dopamine receptors D2 had been implicated in the etiology of the disease. These two scientists used engineered mice to mimic the deficits of short–term memory and attention in schizophrenia. The schizophrenic mind loves to travel; it is jumping from one theme or subject to another unable to focus in the here and now present time. His brain capacity for hallucinations and delusions seem to be a bottomless pit even during his catatonic state. The dopamine receptors D2 suspected of contributing to schizophrenic symptoms are located in two region or groups of cells deep in the brain. These two clusters of brain cells-neurons- are commonly known as striatum, consisting of putamen and caudate. They are part of the basal ganglia which are associated with another chronic and degenerative disease called, Parkinson. Dr. Kandel and his team engineered mice with at least 15 percent more D2 receptors than normal in the striatum. The researchers found some deficits for working memory. A working memory is of a short duration, it is for you to handle the task at hand. It is a here and now memory. Do not bring in synaptic sensitization and permanent anatomical neuronal changes by protein synthesis. The mice in question took much longer than control mice to master the task or game. Control mice were not engineered; their D2 receptors were not touched. The researchers concluded that extra D2 receptors cause permanent brain damage. Dr. Kandel added that damage probably occurs while the brain is still developing. This experiment dealt with schizophrenia symptoms of learning and memory only. It did not deal with hallucinations, delusions, mood and occasional erratic and violent behavior of schizophrenic patients. We will follow up on this experiment as we explore the brain. An interesting observation we like to point out is the connection with the developing brain. There have been other researchers that have followed up attention deficit hyperactive disorder that have turned into full blown schizophrenia. In addition, from parents, teachers and mental health professionals dealing with children behavior, we can observe early symptoms that identifies and later are diagnosed as schizophrenia. During early adolescent hood while the prefrontal cortex is going through immense changes, dendrite growth and maturation may relate to Dr. Kandel observation that the changes he noted occur while the brain is still developing. Dr. Solomon Snyder of John Hopkins Medical Institution in Baltimore concluded that the mouse findings reported in the journal, Neuron could help researchers develop new drugs to prevent such early brain damage from occurring in people who may be susceptible to schizophrenia. Whether I like or not, luminaries like John Nash, Salvador Dali and von Gogh pop up in my mind. They contributed so much to humanity despite their inner pain; we would be a lot poorer without them. Schizophrenia has been an enigmatic disease for millennium. It has been blamed as a punishment from heavens, and/ or a curse from relatives and friends. Schizophrenics have been ostracized, imprisoned, burned at the stake or even guillotined. Most of our medication for schizophrenic is aimed at controlling its chemistry like dopamine and serotonin. The big problem we face is that it is a syndrome rather than a single isolated symptom you can easily identify. The schizophrenic patient goes from a complete flat affect to an emotionally hyperactive individual. During the flat affect he or she may be verbalizing to you a most horrible accident in the family without an ounce of emotion involved in it. During the session you may wonder if the emotional part or parts of the brain were surgically removed from the person you have in front of you. However, the same person seemingly without emotions may become agitated or emotionally aroused by a sensorial stimuli or a past episodic memory and present to you a seemingly normal affect. His mood may change from time to time unpredictably. History records show that we have had the disease with us ever since we began to have symbols describing our behavior.

Alois Alzheimer and spaghetti strings

Before we proceed with our beloved proteins we would like to write down a few words about one of the most dreadful dementia of them all, Alzheimer disease. We mentioned it briefly in another paragraph. Alzheimer disease was first fully described by Alois Alzheimer in the XlX century and continues to be as frightening and elusive as ever. We know that spaghetti string like proteins or peptide chains known as plaques and the protein tau are, as far as we know at present time, are the two main villains or culprit of this degenerative and chronic brain disease. This is the disease that robs you of your personality because it destroys your memories. These plaques block communication between neurons making synapses disappear. A synapse is the space between neurons; a chemical neurotransmitter is released at a synapse. This chemical message will bind to a receptor at a post-synaptic neuron. Another destructive scenario takes place when the neuron internal organelles or fibers break down and form an entanglement that collapses the neuron. The complex and harmonious functions of each neuron cease to exist. What is left of a healthy human being is a body with a head; but the brain inside the head is no longer a chemical and electro genesis machine. The neurons that normally made it possible for you to do complicated and elaborated tango steps, piano playing, do surgery or appreciate Picasso and van Gogh paintings or just take a stroll in Central Park are not available for you anymore. Even when whispering at your ears sweet voices from loves ones, there are not receptors in your brain cells to respond to and reciprocate the feeling. This disease is generally associated with old people, but we have had patients in their forties. It is not an old people only disease after all.

Six years ago, (2006) William Dagust, a physician and neuroscientist said at the time, "Many people believe that we are going to have treatment for Alzheimer in five years." Many miles away in Sweden, a group of researchers tested the spinal fluid for concentration of tau protein and beta amyloid, both strongly implicated in this disease. The group tested consisted of 137 people ages 50 to 86 years old. Those patients, whose spinal fluid at the beginning of the test had abnormally high tau protein and low beta-amyloid concentrations, were nearly 18 times more likely to develop Alzheimer as compared with normal concentrations. Those researchers had hope that there would be effective drug treatment by 2010. There is a lot of confidence in our scientists and technology most recently developed, an effective treatment, in our opinion, is not far away.

Following on proteins

No other molecule in the living world, DNA itself excepted, is chemically, structurally and functionally as complex as proteins. Although the cell nucleus sends out the message for protein synthesis, the message itself, meaning the selection of specific nucleotides that will recognize a complementary molecule with an amino acid, and be pasted together at a ribosome, while other chemical reactions are taking place in the cell cytoplasm organelles and tissues, is an awesome engineering project accomplishment. Crucial among choices to be made (if it is a choice) during this chemical wonder, are codon selections, appropriate amino acids, often within a group of six triplets, amino acid group, carboxyl group, side chain, shape and function as well as size of each protein molecule. How many peptides for a specific protein function? How do proteins respond to multiple internal and external stimuli and still maintain its integrity and function? Your organism makes many protein molecules with highly specialized functions that are vital for the survival of every living cell in our planet. Imagine a protein that exists in a cell or organism in a dry dessert and compare it with a protein in a cell in the coldest region of Siberia or North Pole. Consider a protein that is involved in generating electricity in a brain cell and one in your right foot toe. There are proteins that are turned into highly skilled soldiers of our body like T and B cells of our immune system. I have proteins that are at least over 81 years old and still kicking. Most of my brain neurons are functioning pretty well. Imagine how many hundreds, if not thousands, different types of protein are a doing a specialized job in your body at present time while you read this book. Each one of these proteins has its own particular amino acids sequence. Another educational memory- protein- that we must keep at the tip of our fingers is the property of each amino acid. There are polar and non-polar amino acids that are decisive in protein folding.

From mind speculation to brain research

Just a couple of decades ago, mental health practitioners attributed brain disorders as just mental problems. The issue we want to emphasize here is that mental is considered as having no matter. Mental and psyche was, and still is considered by some quarters as something separate and apart from anything having flesh. By considering mental problems something that we could not assign to a brain tissue or organ, no loci in the brain was considered specifically responsible for any disease in particular. Psychologists ruled out brain tissue and organs as the subject of research and exploration in their efforts to find a reliable and trustworthy cure for problems residing in our head. They reasoned out that if mental problems did not have biological etiology, it had to be caused by environmental factors. They were not too far off in their efforts to find a culprit for psychiatric diseases and disorders. However, they chose the home as the first and most relevant factor leading to brain problems. And, at home, all the blame for all types of behavioral problems was attributed to mama, the mother. Second in line in the culprit list was the father followed by family disputes, poverty, etc. I recall some mental health practitioners blaming the mother as "an emotional refrigerator or an emotional armadillo" for her child's autistic, schizophrenic or schizoid behavior, ignoring a biological base for those brain diseases.

Now-days, scientific approach based on genomic studies, neuroscience as well as all biological related medical diagnostic tools such as PET, FMRI and similar technological medical tools are changing how we look at and treat psychiatric patients. These machines show us that new brain cells are born every day in specific brain regions. Besides, recent scanners show us abnormal connections between neurons, most likely, secondary to a gene mutation. We have identified genes involved in schizophrenia, autism, Alzheimer and other brain diseases and disorders. We know that brain diseases have a biological basis and are not just "mental" conditions that can be cured by a hand tap on the shoulders. We need to go after genes and its twin brother, proteins in its multiple shapes and functions. In above pages we described for you how RNA and amino acids are involved in protein synthesis. You also know that proteins are a major component of every cell in your body; proteins are chains of amino acids. And, you also remember that many amino acids appear in nature, but only twenty are used by our body. Equally important is remembering that shape determines function; and when proteins take the shape and function of an enzyme, enzymes are responsible for many biochemical reactions. So, proteins come in different shapes and sizes from just a few amino acids that will form a chain comprised of hundreds of amino acids. We hope you can visualize how complex this protein synthesis is. Let us imagine a protein composed of a thousand amino acids all packed tightly so it does not get entangled with other amino acids forming another protein or other intracellular subunits. It seems that the job that lies ahead of us is not a piece of pie. The beauty of the job that lies ahead of us is that we love challenges. Working on proteins that are particularly relevant to medical research and the solution of general and psychiatric diseases I saw incentive we cannot fail to celebrate. You can help by attending seminars and workshops provided by researchers throughout our country.

Our basic carbon chips

Recounting our story, you and I are basically made of carbon chips as a basic functional unit of multiple cells. Carbon, oxygen nitrogen and hydrogen comprise about 95.6 percent of all the elements our body needs to survive. A carbon atom is a very dynamic atom; it makes many decisive bonds with atoms and molecules of other elements that are necessary for all living things in our planet. Carbon atoms are involved in amino acids molecules. All amino acids molecules must have a carboxyl group and amino group, and both groups are joined to a carbon atom. The carbon atom that these two groups of molecules are joined to is known as a-carbon. You know that the basic unit of your entire body, whether it is a hair on your head, a toe nail on your left foot or the upper lip of your mouth, it is made of proteins. And... proteins are long chains of peptides that in turn are made of amino acids. These protein chains coiled up as if it were a ball and therefore are known as globular proteins. So, proteins are polymers of amino acids joined head to tail in a chain that could be several hundred units long. It is folded in a three dimension structure. The union or bond between two nearby amino acids is known as a peptide bond. The carbon atom is always involved in everything that is cooking in your body. Just think about it, of the twenty types of amino acids found in all proteins, no matter which amino acid is chosen to form a polypeptide chain; it must have a carboxyl group at one end- known as c-terminus and amino group at the other end, known as n- terminus.

The beauty of all amino acids is that they repeat over and over again in all proteins. No matter what part of your body you begin to take apart for analysis, you will come across with some of the same old good amino acids. You may play around with it adding growth or degrading factors, you will end up with amino acids. The funny part of this game is that you will be learning about yourself. Even before you opened this book, the image of DNA and RNA came to mind. These two acids plus carbon atoms, proteins and amino acids are all about you. Later on, you became acquainted with the famous letters making up the two famous macro-molecules of the two above mentioned acids. The letters we are referring to are: t-a, c-g for DNA and a-u, c-g for RNA. Please, remember these different sets of letters for these molecules.

We have also use the word nucleotide quite loosely, but what is a nucleotide besides the usual connotation of a base? A nucleotide is in itself, a molecule, which in turn, is itself made of atoms. More carefully defined, a nucleotide is a molecule formed of nitrogen containing ring compound that is joined to five –carbon sugar. Basically, there are two types of nucleotides that we like to work on. Nucleotides that contain ribose are known as ribo-nucleotide (RNA). Nucleotides that contain deoxy-ribose are known as deoxy-ribonucleotides. These are long words that soon became shortened in our daily conversation. In general terms nucleotides are referred as, or named after one of the bases it carries. That is why you read a thymine nucleotide although it is implied that it carries one or more phosphate group. Yes, this is the phosphate- sugar backbone that Watson, Crick, Wilkins and Rosalind Franklin were struggling with while trying to come up with the double helix structure. Were the bases best hanging outside the backbone, or inside making hydrogen bonds? One nucleotide you will not be able to ignore while studying the two great molecules, DNA, RNA nucleotides is adenosine triphosphate or ATP for short. It is an important energy carrier molecule involved in many chemical reactions.

From transcription to proteins

Messenger RNA can leave the cell nucleus as a single strand and travel through the cytoplasm and land in a ribosome. This protein assembly factory also receives another RNA single strand known as transference RNA. This molecule reads the message- three letters- that the messenger RNA clearly displays and the ribosome pastes it together. After it is pasted together by the tiny globular molecule ribosome, it is kicked out as a peptide; multiple peptides form a long chain that we know as proteins. A messenger RNA lands in a ribosome if another RNA molecule does not interfere along the way from the cell nucleus to a ribosome in the cytoplasm. In terms of space, time and size, it is a long distance to go from one place to another, especially, when it has to go through many smaller organelles busy each doing its assigned job. Besides, there is another RNA molecule known as interference RNA that bonds to target sites on a messenger RNA molecule altering its shape and function. The vast majority of us are blessed from the very beginning of conception if we ponder on the complexity of life as it grows from a zygote to a fully developed and healthy human being. It is not just i RNA that may interfere with the formation of proteins, but also micro RNA may do it too. The first micro RNA gene was discovered in 1993, five years later, RNA interference was discovered. This is the RNA molecule made me grow goose pimple when I began to understand its potential power and intervention in protein synthesis and gene activation and suppression. Nicholas Wade wrote for the New York Times on 6/21/05 that i RNA is "a system for silencing genes by tricking the cell into destroying the gene's messenger RNA before it can generate its protein product...Together with transcription factors, micro RNAs may play a role in cell differentiation, the formation of many specialized types of human cell from a single generic type...Micro RNAs create an environment, also tailored for cell type, in which some kinds of messenger RNA can flourish and others are diminished or repressed." We are trying to create for you a picture of the role ribonucleic acid plays in protein synthesis. We had high expectations on our ability to make research tools and drugs from distinct from of RNA molecules. However, RNA has proven a hard bone to crack as above mentioned companies have found out. We are dealing with the oldest bio-acid we know of. When DNA showed up, RNA was an old man with grey hair. In the same paper and same date as above, Brian Libby wrote: Dr. Kenton Gregory at St. Vincent Medical Center at Portland is one of a handful of researchers working to create replacement tissue from a naturally occurring protein, elastin. Elastin is known as a matrix protein in that it holds the cells together into tissues and provides a natural support and flexibility. It gives skin and blood vessels their elasticity... Another matrix protein is collagen which is for the tensile strength of tissues.

Proteins, memories and Eric Kandel

Eric K. Kandel brilliant work and discovery of anatomical changes taking place at synaptic sites when permanent long term memory is established in our brain is awesome and inspirational for many of us trying to understand ourselves. To our delight, once more proteins synthesis takes first place with all the complexity of transcription, translation and peptide formation already discussed on above pages.

Commenting on a discovery by Louis Flexner at the University of Pennsylvania that drugs that inhibit the synthesis of proteins in the brain, disrupt long term memory if given during and shortly after learning, but they do not disrupt short term memory. This finding suggests that long-term memory storage requires the synthesis of new proteins. Now, a very interesting and crucial question arose among all of us. Where does the new protein grow in a brain cell during the formation of long term memory? Does it grow on the same site as short term memory? Is long term memory an outgrowth of short-term memory? Is there a specific region or cluster of neurons in the brain responsible for the formation of proteins for long term memory? Doctor E. Kandel has the answer to our questions. His research at Columbia University provided him with a live tool he used it wisely and fruitfully. He wrote: "Short term memory produces a change in the function of a synapse, strengthening or weakening preexisting connections; long- term memory requires anatomical changes. Repeated sensitization training causes neurons to grow new terminals giving rise to long term memory." Neuronal connections already exist, you activate them through a learning experience, let's say when you learned the first three table of multiplication. You enjoyed it and kept repeating it several times, and even went to the next step of learning the six table of multiplication. You felt great; you enjoyed the "feeling good" that we experience when we accomplish something that satisfies us and the love ones around us. You go out and play ball, play with your dog and even watch television for a short while, and you can easily, without any effort, recall all the multiplying tables you have in your brain as long term memory. This type of memory requires protein synthesis, proteins that will be attached to or seen as a protrusion coming out of dendrites. It is as if a new small branch sprouted out a neuron. This is what E. Kandell called anatomical changes in brain neurons. A note of clarification, this is happening in neurons. The brain also has glia cells, but at present time, its role in memory formation is not clear yet.

Another teaching lesson from Kandel is the following: "The plasticity of the nervous system- the ability of the nerve cells to change the strength and even the number of synapses- is the mechanism underlying learning and long term memory."[34] Among memories, there is one I never forget when first I entered the physics classroom, E= MC2. Albert Einstein was such a mountain figure that the professor at the University of Puerto Rico had his picture covering the main door as we entered his classroom.

The learning process has been the subject of study, debate and speculation for millennium. Methods of teaching have been developed by many outstanding educators throughout the ages; some methods have survived the challenges of time, place and needs of society with slight revisions while most lagged behind and disappeared. How to pass on our body of knowledge -our wisdom- to the next generation is a great challenge to our society today. Equally challenging is the content of the body of knowledge we want our children, and our children's children learn in order to survive in a highly competitive society. In the year 399 BCE in Athens, Greece, the greatest philosopher the western world has known; Socrates was forced to drink the hemlock, a poisonous herb; he was accused of perverting Athens youth by unorthodox methods. The debate is even more challenging and controversial than ever before. We have issues like creation and evolution, science and humanities including art and history. Some individuals are advocating reducing the last two to a very minimum of time while science and technology should be raised to first priority. Is education a State responsibility or should it rest within the family primarily? Should education of our children be left to educators and parents, or should politicians be in charge of our schools system as they are in charge of legislating for all of us? The XX century left many scars on us under Nazis and Communist regimes. We are

_____
[34] - Eric K. Kandel, op cit, p. 218.

learning rapidly how learning and memory take place in our brains. Sensitization and permanent neuronal structural changes can be easily manipulated in the classroom. These are issues that need our attention before it is imposed on us.

More on memories and learning

Once we had grasped the idea on how learning and memories get permanency in our brain cells, we could not stop searching how Dr. Kandel was engineering his exceptional manageable research toy, the aplyssia. We were familiar with short term memory also known to many of us as working memory. We were familiar with H.M. surgery on both temporal lobes and removal of most of the hippocampus. After surgery H.M, could not store memory for long term use. All his memories after surgery only lasted a very short time. Somehow, functional memories could not find their way to other brain cells for storage and future use. H.M could not transfer having eaten a sirloin steak with baked potatoes, French salad and a glass of wine to a cluster of neurons for permanent storage. Soon after he finished his delicious dinner and walked into the crowd on the street, you may want to ask him how he enjoyed his steak; he would simply replied, what stake? The steak dinner was inside his stomach, but his brain did not register it. You may wonder if brain cells around the hippocampus in the temporal lobe could be generous enough under the plasticity theory and take over memory storage for H.M. We know that brain cells often take over functions not originally theirs. The physician that took care of H.M. health for many years reported that he never recovered his past memories after surgery. Related to the hippocampus, Laura Beil writing for Science News said: "The hippocampus encodes and prepares new memories for storage, then dispatches them to different parts of the brain. In 1989 scientists reported evidence that the human hippocampus is not only a depot for memories, but also, a birthplace for neurons-thousands each month." The nursery for nerve cells is restricted to a raisin-size region of the hippocampus called the dentate gyrus. Supporting the above statements, Fred Gage, a neuroscientist at the Salt Institute for Biological Studies in La Jolla California said that at any given time, about 3 to 5 percent of the cells in the dentate gyrus are in some stage of growth.[35] He further added that most of the dentate gyrus is formed after birth; a lot is formed during the first four years of life. This is very important to know as it relates to growing up children during a critical period for learning and forming new memories. Parents and educators, specially, early childhood educators, should take advantage of this information to plan teaching lessons because we are talking about one of the main region of the brain engaged in learning and memory formation. These are memories that will mold our personality for life. For instance, the earlier you expose a child to another language or languages the better. The same is true for mathematics, science, astronomy, etc. Neuronal circuit and pathways as well as synapses are established and strengthened for adult life challenges.

Memories, brain and learning

Whether we like or not, we are, for the moment, stuck with proteins. You know that your wedding date was embedded in your brain as a protein. Therefore, we will continue our inquiry into proteins and memory formation. I carry toxic episodic memories in my brain that pop-up in my mind without an invitation. Likewise, I carry sweet and educational memories that make my life happy. Our interest on protein synthesis, memories and disease is not just an intellectual exercise; it is very personal one. Besides inheriting thalassemia traits from my parents, I am short sighted, and when I come out of a dark room and face light I become almost totally blind for a few seconds. Of course, I wear corrective lenses. I am not an

---

[35] - Science News, Vol. 179, # 3, 01/29/2011, p. 23-25

albino person. I have been examined by an ophthalmologist and he told me that I lack some vitamins and tissue on my eyes that sun rays blind me. My eyes are slow adjusting from darkness to light. Some researchers have linked it to the pineal gland deep in the brain. This tiny cluster of cells is responsible for releasing melatonin and is also involved in the wake –sleep cycle. It seems that probing my brain is not a curiosity at all, but a necessity in discovering myself through my strength and liabilities.

We have repeated over and over again that DNA molecule nucleus is the command center for everything that takes place in the cell. However, this seems to be partially true. For example, strengthening or weakening a synapse is more like a localized issue than a command from the cell nucleus. For instance, if I do not repeat the order and names of planets in our solar system several times and read it aloud, it will vanish from my brain. Someone whose name I do not remember said: "Use it or lose it." Our brain has trillions of possible connection among dendrites. It can store many things, but it is my belief that the brain does not waste energy in trivial things that do not use regularly. The following might not be a good example, but I will write it anyhow. Coming from my home in the countryside I rented a room where an elevated train passed by every half hour. The first nights were torturing nights plucking my ears with strips of cotton trying to go to sleep. Within six months, I did not see nor heard the train passing by anymore. The cotton in my ears, the train and the noise had gone forever. Yes, it is call habituation. My brain has survived thousands of years, and I hope it can survive many more. My brain uses synapses as needed; it also closes synapses not in use. It is the law of conservation of energy.

CREB protein

We had read about protein kinase A traveling to the cell nucleus for the purpose of activating a gene or genes. In addition, we had learned from Paris that Jacob and Monod, the E. Coli experts, had discovered that messages from a cell surroundings can turn on genes regulatory proteins. We were assuming that a particular place protein needed in a particular place was synthesized by the relevant gene or genes. However, in all honesty, E Coli has never been in my list of research models. Just to think that I did not invite it into my intestines, share my food; and even create many unwanted symptoms on me, make me grow goose pimples. Perhaps it was my prejudice about these little parasites that prevented me from learning more about Jacob and Monod`s work. However, Watson book on DNA and E Kandel book in Search of Memory made good use of E Coli secrets as revealed by those two researchers at the Pasteur Institute. E. Kandel came across a cycle response element-binding protein (CREB) for short. Creb is a gene regulatory protein involved in many processes including learning and memory. It binds to a promoter in a gene. In his laboratory he found that if he blocked the action of CREB in the nucleus of a sensory neuron, he prevented long-term, but not short term memory. It means to me that he was blocking signals to the central command center, the nucleus. Therefore, there were no orders for transcription to take place; consequently, there were no proteins in progression. Kandel wrote in his book: "Blocking this one regulatory protein blocked the entire process of long-term synaptic change!" The activation of CREB leads to the expression of genes that change the functions and the structure of cells. Creb proteins come, at least, in two forms, one for activation and another for suppression. This is the on- off switch nowadays. For us, it was not a surprise that protein synthesis takes place in the cell body; but Oswald Steward from the University of California had shown us that protein synthesis also take place at synapses; it was quite confusing for us. It was confusing because it reverses the whole functional process of a neuron. It bypassed electro-genesis, hillock and the neuron nucleus itself. We will close E. Kandel in this chapter with the following: "The process initiates long-term synaptic facilitation by sending protein Kinase A to the nucleus to activate CREB, thereby turning on the effectors genes that encode the protein needed for the growth of new synaptic connections. The other process perpetuates memory storage by maintaining the newly grown synaptic terminals, a mechanism that requires local protein synthesis."[36]

I will add a final comment on CREB proteins and my companions, the micro RNAs. Among micro RNA there is a mi R 212 which is shaped like a hair pin. In experiments done with rats that were engineered to consume cocaine, the rats decrease their addiction when micro 212 increased. Perhaps this micro RNA can be translated into a therapeutic tool for human addiction. Further experiments demonstrated that blocking the formation of miR212 increased the appetite for the drug. At present time, we seem to have good command over RNA in its multiple forms. We can clone it, synthesize it and engineer it at will. This tiny RNA molecule volume is related to a CREB, a protein involved in long term memory and anatomical changes in neurons. Eric Kandel tells us that protein Kinase A goes to the nucleus of the neuron to activate CREB which in turn, turns on the gene that codifies the protein needed for growth of new synaptic connections. We beg you to bear with us our curiosity for ribonucleic acid. Whenever we turn a page in any book of biology and all its processes there we find RNA involved in our lives. We focus our eyes and attention on tools that may help ameliorate the pain of our brothers and sisters worldwide. The pain outweighs the benefit, if any, when cocaine or heroin hijacks the brain of a person. "Protection against cocaine addiction may be a side benefit of miR normal job of regulating CREB production and other biochemical processes in the brain, said neuro-scientist, Paul Kenny."[37]

---

[36] -E. Kandel, op cit, p, 270
[37] - Science News, Vol. 178, #3, 07/31/2010, p. 11

Two Nobel Laureates advocating for RNA

A 1989 Nobel Laureate on chemistry for the discovery of RNA multiple functions, and in particular, ribozymes, wrote for Nobel Prize Org in 2004 that he, Thomas R. Cech and Sidney Altman, independently of each other, found that RNA could fold into complex shapes and catalyze biochemical reactions, a function previously thought to be restricted to protein enzymes. Thus, RNA was sometimes an active participant in the chemistry of life, not just a passive messenger. He added: RNA often controls the expression of genes, another role that had been thought to be at least mostly the domain of proteins called repressors and transcription factors." The discovery of ribosomes as a ribonucleic acid molecule that assembles protein just added scientific recognition of this acid vital- crucial- role in chemistry, development and evolution of life on earth. It was seen as home- run for RNA advocates claim that it has existed before the DNA world. RNA could replicate itself and multiply while DNA needs RNA molecule for the synthesis of life building blocks, proteins. One impressive feature of ribosomes is that it decodes a very large number of messenger RNA to make protein for each organ, tissue and function in our body. Imagine ribosomes putting together all the units from a messenger and transfer molecule coming out of the cell nucleus to make a protein for a specific cell in the prefrontal cortex, or the heart, or liver or a muscle in my left leg. This is, by no means, a small job even for a summa cum laude. Remember, ribosomes are not located within the membrane of the nucleus to be supervised like grandma used to; ribosomes are located figuratively, miles away in the cytoplasm.

According to the above named researchers, a ribosome is composed of three RNA molecules (four in some species) and dozens of proteins. Most recent pictures of ribosomes at atomic level show that it is in fact composed of RNA with no proteins in the vicinity; it is indisputable evidence that indeed it is an RNA catalyst. It is in charge directing mRNA and tRNA interaction in the process of protein assembly. At the same time that these discoveries were taking place, catalyst and enzymes were almost identical. Enzymes were the catalyst agent per excellence; RNA was far behind, a passive agent you needed no to bother with. However, during the last decade of the XX century, and most recent discoveries, has dramatically changed our concept about this very old ribonucleic acid, its multiple shapes and functions. During the birth of the double helix in the early 1950s there were hardly any books on RNA if at all, almost hidden behind piles of illustration of DNA and the double helix structure.

The third millennium brought us a new experience with our grandpa, the RNA molecule. We have seen how it turns gene expression on and off and plays a definite role in the making of proteins as well as chemical reactions within the cell. The beauty of this technological advancement in many areas of science is a blessing for humanity. The French and American revolution of the XV111 century opened the door for our little genie that was kept imprisoned inside our skull. R. Descartes "cogito ergo sum" could no longer be kept a prisoner of the pineal gland deep in the brain. The French philosophers of late 18 century knocked down many walls and the American mind built a new and democratic society. The human mind was set free; it could not any longer be kept enslaved in his body by an antiquated and obsolete body of knowledge. We challenged authority handed down to us by the force of the sword, prejudice and tradition that benefited and protected the few at the expense of the majority of humanity. Benjamin Franklin and Thomas Jefferson, although part of the American elite, their mind was not constrained by society rule of behavior at the time. Franklin was a man of science ahead of his time and; on Jefferson, we are still debating his thought and actions. They were challenged by equally brilliant compatriots. We became the free man of planet earth. Even the most repressive regimes use the word Democracy in reference to our system of government.

A pioneer neuro-scientist, Miguel L. Nicolelis, wrote that freeing the brain from the limits of our terrestrial bodies will allow the paralyzed to walk. Through this liberation of the human brain from the physical constrains imposed by the body, the disabled may rise from wheelchairs. Humans have shown that brain can be directly linked to machines in a laboratory setting. He adds: "A few years ago my group demonstrated the feasibility of linking living brain tissue to a variety of artificial tools. It will require a new generation of high density micro-electrodes that can be safely implanted in the human brain and provide reliable, long-term simultaneous recordings of electrical activity of thousands of neurons distributed across multiple brain locations. Doctor Nicolelis envisions a bidirectional, thought-driven interfaces, operating a myriad of nanotools that will serve as our new eyes, ears and hands. [38] Similar techniques like DBS or deep brain stimulation are used today to assist Parkinson patients gain control of their body, among other things, controlling tremors and bodily balance.

Previously unknown RNA functions

Among its previous unknown functions are as follow: "It can provoke- cause- termination of transmission of messenger RNA, it can interfere with ribosomes from translating mRNA or it can cleavage (split or brake up) mRNA into destruction. The ribonucleic acid molecule has a cleavage or splitting mechanism that seems to work even in itself. RNA existed at the side of DNA for most of the second half of the XX century, but recently, with so many shapes and functions of RNA molecules; it seems that, along with epigenetics, a new branch of molecular biology is needed. We are referring to RNA not only as a research tool, but primarily, as a medical

---

[38] -Scientific American, 02/2011, p.81-83.

drug treatment modality for many chronic and degenerative diseases that we are struggling to conquer and stop the suffering of millions of people. RNA as a gene switch and protein most prominent assembly agent and breakdown facilitator seems to have a very promising role in medicine in the very near future. Another attribute of RNA not previously suspected besides a formidable catalyst agent is its ability to come out as a double strand. Normally, RNA is a single strand, meaning that only one copy is transcribed from DNA. However, given that it is a single strand with no partner to link with, it possesses the intrinsic attribute of folding back into itself and/ or takes any shape necessary for a function. You can see RNA hair pin that are, functionally, double RNA strand. So, a single strand of RNA can fold or coil itself in different shapes. For the sake of clarification, you can see it as a long spaghetti strand, a hair pin or pair of pliers, a bead of a necklace or rosary beads or a circle like structure. It attaches itself to, or allows other molecules to link with as on histones in the chromatin. The linkage attribute takes many forms. In addition, you have the well-known RNA interference. This versatile molecule can not only block protein synthesis, but with the assistance of enzymes like Dicer and Slicer, can cut double strand RNA into 20 base pair fragments. One of these two strands can be transported to a matching sequence of a messenger RNA for protein formation while the other strand may face degradation, meaning destruction. By now, you are probably almost convinced about the crucial role of RNA on our life. But this short story about RNA does not end in here. It is not just protein making and gene function interference.

And… talking about meddling on genes affairs, you have not forgotten RNA interference; it has become a very important technology for scientists to target gene inactivation in their never ending search for drugs for the treatment of genetic derived illness. Suppressing a gene may lead you to discover the causal factors of a disease or a syndrome leading to a disease in the brain or any other organ of your body. An

RNA molecule that has the power to block the construction of the basic building parts of my body not only during the translation process, but also, inactivating my genes or your genes, is a powerful cluster of atoms formed into a molecule that we need to understand and gain to our side. Please, bear with us, because we consider your brain is extremely important for us to ignore just a tiny bit of it. Your protein factory, the ribosome from all species consists of two subunits: a small subunit that decodes messenger RNA and a large subunit that catalysis peptide bond formation between the growing polypeptide chain and each new amino acid. This is part of the process of translation we are committed to fully understand in eukaryote ribosomes which is much more complex and highly regulated than non-eukaryote ribosomes.[39]

A little bit of history

A little bit more of history of this ribonucleic acid does not hurt anyone. It began over a decade ago here in U. S.A., Europe and Japan basically. Just as the race among scientists for the search of codons or triplets, amino acids, and protein synthesis got researchers on both sides of the Atlantic busy in their laboratories; the RNA molecule many roles and shapes put in action many biologists on the race to decipher this molecular acid function and shape. The discoverers of the double helix in England were interested in this molecule that appeared to be engaged in all aspect of our life. Initially, we were happy reading and studying DNA double strand and RNA single strands such as messenger RNA and transference involved in protein synthesis. But as more functions were showing up, two young scientists: Craig Mello and Andy Fire had not overlooked the power of RNA molecule to intervene in the process of protein formation. Both, working in a little parasite that loves to live in mammals intestines, including humans, E. Coli, provided the first ever demonstration that

---

[39] -Science, 02/11/2011,Vol.331, p.681.

double stranded RNA could interfere with genes and even silence it. Do not forget this statement; it is very important. Once again, the molecule RNA that is so important and indispensable during the synthesis of proteins, can intervene during the whole process of making proteins for your body and mine. Among other things, it can bind to a messenger RNA that comes from the DNA nucleus and derail its synthesis; but, in addition, do not be alarmed, it can also silence genes. Mello and Fire posed a huge challenge to biologists. To think that a ribonucleic acid molecule can alter and even stop the transcription process, a domain function of the Powerful DNA molecule, was heretical. Ever since I first read an article on this interfering molecule, I began to wonder what RNA got to gain when it decided to be the DNA molecule workhorse. I became interested in the RNA world before the DNA- protein world. Within each cell we have three wise and powerful molecules working together to build, develop and maintain a very complex organism like yours and mine. The three king molecules we are referring to are: DNA, RNA and mitochondria. Each one has definite and clear roles to play; and the have played it very wisely and beautifully. When I consider that silencing RNA molecule can be used to mark specific DNA sequences for deletion, and guide modification of chromatin structure, I get goose pimples. However, I try to bring relief to myself by saying to myself this molecule has worked remarkably well on me. The mitochondria with its ATP molecule cannot be counted out during the balance of power among the three rulers of our chemical factory, our cell. A recently published scientific journal (2011), tell us of a team of researchers that made use of mice mitochondria DNA linked to health benefits. In mice, a regime of endurance training-running for 45 minutes three times a week over five months induced mitochondria bio-genesis, increased mitochondria respiratory capacity, and prevented mitochondria damage."[40]

---

[40] -Science, Vol. 331, 03/04/ 2011, p. 1115.

An established fact, an RNA world

An RNA world existing before a protein coding one, which is the DNA world, is an established fact among modern scientists. The irony concerning this scientific truth is that it became accepted after the discovery of the double helix. Ironically, it provoked us into laughter and further self-questioning about its functions. The seemingly paradox that we are trying to deal with is that multiple cell complex organism developed after the protein coding DNA world took over. However, the work horse in protein synthesis is the RNA molecule in its multiple forms. We have said that the DNA molecule incorporated or swallowed in RNA to do its job in the process of forming or constructing proteins. Similarly, we also pointed out that the wise DNA molecule incorporated the mitochondria. We have not been able to resolve this issue to our satisfaction. However, we are very happy with our progress understanding both macromolecules and the sequencing of the human genome. We would like to share with you an excellent article written by Eric S. Lander in the scientific journal, Nature. We will write down just a few things, although incomplete, but it sounded very interesting to us. The sequence of the human genome has dramatically accelerated biomedical research... and our understanding of the biological functions encoded in the genome. The greatest impact of genomic has been the ability to investigate biological phenomena in a comprehensive unbiased, hypothesis free manner. In basic biology, it has reshaped our view of genome physiology, including the role of protein-coding genes, non-coding RNA and regulatory sequences. In medicine, genomics has provided the first systematic approaches to discover the genes and cellular pathways underlying disease. He further added the discoveries of location where DNA and chromatin are modified as well as the study of inherited variations or somatic mutations. Sequencing is also applied to RNA transcripts and its multiple forms. We have also clarified the volume of protein-coding genes in the human genome as consisting of 21,000 genes. At the beginning of the genome sequence project some had put the number of protein-coding genes as high as 150,000. On small non-coding RNA, Eric S. Lander said that a few dozen mi RNA have been shown to have key regulatory roles; recently; a new class of small RNA called Piwi interacting RNAs, has been discovered, they act to silence transposons in the germline. Transposons may be seen as drivers of evolutionary innovations. In addition, we have benefited from sequencing the genome, the identification, location and roles played by epigenetic markers. The vast majority of human variants have been discovered, and they are under study. In its clinical application, DNA sequencing is being increasingly used to assign patients with an unclear diagnosis to a known disease. In psychiatric diseases, genomic studies have identified common variants in bipolar disorder and schizophrenia and rare deletions in autism.[41] We could not provide a complete account of the article, but we hope you can get access to the journal and enjoy as we did.

Dr.Tom Misteli from the NCI and DNA

Ten years ago, a blueprint for a human being, the complete list of the DNA now famous letters ATCG was announced to the world. In celebrating a decade of genome research, Tom Misteli, a senior investigator from the National Cancer Institute in Bethesda, Md. said: "Biologist have known for long that DNA chromosome folds up in complex ways. They have now demonstrated that individual chromosomes occupy distinct territories in the nucleus and that some chromosomes prefer the nucleus periphery, whereas others like to cluster close to the core. Moreover, where chromosome resides, and which chromosome lie near one another, can strongly influence how cells function." (We did not fail to notice the word –strongly- to define the influence on the cell.)

---

[41] -Nature, Vol. 470, #7333, 02/10/2011, p.187-197.

Tom further added: Where chromosomes lives seem to influence whether the genes it carries are turned on or off."[42] A gene gets turned on after proteins known to us as transcription factors get together on regulatory regions of the gene. Our workhorse, the RNA, this time in the form of protein RNA polymerase, comes in to transcribe the gene's DNA famous four letters into RNA multiple copies you are already familiar with. The DNA you will find it in the nucleus, while RNA is always moving around and changing its shape and function. Following Professor Mysteries observations, he generalized and said that researchers now know that the nuclear periphery has silencing effects on genes, and the center promotes activation. He further elucidates the intrinsic working processes taking place during protein synthesis adding that hundreds of genes that encode ribosomal RNAs are transcribed together in the nucleolus- a nuclear substructure large enough to see under a microscope. [43] The nucleolus resides within the nucleus, but has its own membrane. Do not look for it around ribosomes or nearby organelles. Within the cell there is a good busy traffic going on from one area of the cell to another. You will find organelles transporting subunits from the cell basic components, from one station within the cell to another. You will also see polymers brought into the cell through its membrane as well as sharing or transporting material outside to neighboring cells. If your organism has been invaded by unwelcomed bacteria and virus; please, use your imagination for the traffic jam that will take place in millions upon millions of cells in your body. Your T and B cells with an army of combatants will order an attack to eliminate toxic invaders.

Micro RNA regulatory role

The celebration a decade of genome sequencing cannot ignore micro-RNA regulatory role in gene expression. "Over

---

[42] - Scientific America, 01/ 2011,p. 70-73
[43] Scientific American, 02,2011, p.71

the past decade, small RNA emerged as a new class of key regulators of eukaryotic biology. This diverse class of RNAs includes small interfering RNA, micro-RNA and PIWI interacting RNA, all of which associate with multiple protein components within a complex to regulate partially or perfectly complementary transcripts. Among the objectives the researchers were trying to understand is how mi RNAs bound to Argonate proteins, bring about gene silencing. There is abundant literature indicating that translational inhibition and m RNA decay are coupled throughout biology. [44]

Unbelievably naïve

Our intention during this essay has been somewhat naive, to analyze and understand the code of life Watson and Crick began to play with during the middle of the 20 century. We began with double helix and proceed to nucleotides, amino acids and protein formation, are among the most interesting stops we found in our tree of life. We stopped for a while at the intriguing RNA molecule and its many curious and useful conformations and functions it takes to serve us best. Of course, we could not bypass epigenetics and multiple DNA mutations and histones modifications which were briefly touched on. Perhaps our intention during this trip, although unconscious, there is the very simple and naïve question of my origin as I am now. The nagging unknown of my original birth on planet earth millions of years ago I will not be able to solve, but you will. My own ancestors including my grandparents did not have this mind bugging urge to explain the unknown and relied upon generations of old tales often written in their sacred book, the Bible. More recently, our origin has been attributed to aliens or meteorites that came from another solar system. Both, my ancestors and recent theorists are not too far apart, both place my origin outside planet earth. However, for some reason, we tend to stick with the "chemical soup" theory of

---

[44] - Science, Vol.331, 02/04/2011, p. 550-53.

mother earth. This wonder soup, under certain pressure and heat, attracted selective atoms and molecules that in due time became living cells. It has been a long, long time ago that in our earth history that this spark of life took place, if at all. Most recently, science fiction and some respectable physicists and astronomers are advancing a theory that aliens with super-intelligence may have visited earth earlier in time and we are just guinea pigs under constant scrutiny and observation. It looks as if we are laboratories to watch and see how we behave in certain conditions; how we take care of ourselves, and ultimately, take care of planet earth.

Here and now with amino acids and proteins

All the above theories sound very interesting, but we prefer to stay close to mother earth. Instead of traveling out of our solar system in search of myself, I am stuck with amino acids, peptides and earth elements that constitute life in here and now. Of course, the double helix and the gene code is a giant step forward.

But, when did life begin in this beautiful baby? Mom rightfully would say, "This is my baby, I gave birth to him." Well, you pass judgment over moms claim. Did life begin at conception? Or should we say during the first cell division, or three days later during the morula phase of embryonic development with an internal mass of eight cells? Someone might argue that we must wait another three more days until the phase of blastocyst take place and cells begin their journey to specification. Perhaps, a wiser man join the conversation and argue: "Life does not begin until cell specialization has begun with cells for a heart, brain, lungs, liver and the remaining organs and tissues of our body are in place." You are right, now we are not talking about the same thing. The original unknown was left far behind, but this is what is happening during most of the meetings I have attended to on this subject. My inquisitor may interrupt me and say, Oh, but this is democracy, my friend! Of course, it is democracy but, we are not talking about democracy; besides, democracy in itself is in need of a new definition.

St. Agustin and Darwin join our group

Suppose that our group of science curious individuals is joined by someone like Socrates arguing in favor of his soul? Even a better yet scenario would be Aristotle, St. Agustin, Darwin, Einstein and Oppenheimer joining our lively group. It seems to me that in this hypothetical case, we have a centrifugal chemical soup that, nevertheless, could result in an implosion provoking a tentative truth. But, sadly to say, whatever truth the implosion group may come out with, it is still an educated guess of a group of people in a bit of dust in an infinite multiverse. Please, do not blame me for beginning this mind torturing dialogue; I blame the people I named above including Watson and Crick. In the meantime, we will continue our journey without more outside interruption. Brain growth and differentiation comes under control of genes around five to six weeks after conception in humans. However, this pre-natal phase of development is very critical for the fetus health. His mother diet and life style and environmental factors will influence the fetus development long after he or she leaves the comfort and protection of the mother's womb. We are all aware of cigarettes smoking, drug abuse, biological predisposition, hunger, physical abuse and chronic stress during pregnancy.

An estimated 100 billion neurons and around 800 billion glia- cells need the correct amount of vitamins and minerals to make the appropriate and corresponding shape, traveling within the cell, building and discharging neurotransmitters for its final normal function. A deficiency in the basic elements necessary to build amino acids that ultimately will form proteins for our body is well established around the world. Following each step during the fetus development is not the best thing to do with our present technology, but we have our four legged friend, the mouse, and other animals at our comfort and convenience in our laboratory.

Stress and ADHD in pregnant mothers

Let us take stress as a subject of study. In a pregnant mouse, the exposure of the future mother stress hormones can lead to anxious behavior and hyperactivity in the offspring. A relatively recent longitudinal study of over 7,000 mothers and babies (humans, not mice) at Imperial College in London concluded that maternal stress may account for up to 15 percent of diagnoses of attention deficit hyperactive disorder. [45] During the first decade of life, boys and girls spend much of their time playing together often under loving parental supervision and control. During this period of human development the central nervous system is busy making many necessary circuits and pathways. During adolescent hood, not only sexual hormones are having a party; but glia cells and neurons dendrites and spines are trimmed to serve specific purposes corresponding to biological, physiological and behavioral stage of development going on during this critical period of human growth. You have association neurons, motor neurons, visual, acoustic, limbic system, and many others brain cells making appropriate adjustment to best serve our organism.

The prefrontal cortex, the brain area par excellence for cognition and high level decision making is the last neocortex area or region to complete maturation. Unfortunately, but understandingly, this is the time when we feel that we have to experiment with everything around us that adult people take for granted. We experiment with drinking alcohol, smoking and car driving. Car accidents resulting in injuries and death are higher during this period of our life, unfortunately, while our brain decision making cluster of cells are still in the process of making corresponding axonal connections. It does not mean

---

[45] -New Scientist, 04/04/2009, p.28

that connections are permanently closed; synapses are busy receiving and sending out chemical messages with potential new pathways. Adding to adolescent hood turmoil, a dreadful brain disease, schizophrenia, takes its toll during late teens and early twenties. Brain areas like nucleus accumbens and ventral tegmental area are easily recruited by our reward system that ends up hijacking the whole brain, and addiction takes over the person's life. According to reliable studies in different laboratories, once an addiction drug has hijacked the brain, it provokes permanent changes to those areas. The addicted individual exists to sustain an addiction that has taken over part of his brain that dictates his behavior at all cost. The addicted individual does not enjoy his drug because the state of feeling as you and I experience it; is no longer part of a normal brain behavior.

Day dreaming

Our hypothetical group that St. Agustin and Darwin joined us to help to come to an educated guess is based on many TV documentaries that have attempted to solve some of our problems. The groups have ended, in the majority of cases, just like in the beginning, each member holding to his or her beliefs. So, we were not dreaming, we were, I guess, mind traveling for a short time. But, is day dreaming a resting period of the brain? Day dreaming has had multiple interpretations and meanings for many people around the world. Some researchers have concluded that it allows brain memories to be analyzed and organized into content related compartments in brain cells. Occasionally, researchers and psychologists have associated daydreaming with neuron affinity, circuits and pathways compatibility and normal functioning. Much less reliable are philosophical and religiously inclined individuals that postulate that daydreaming is the time for the soul to leave the body momentarily to communicate with its superiors, whatever it might be. All the above defend each other position with about equal tenacity and conviction. Unsurprised, when I began to read an article on daydreaming that related to brain diseases like schizophrenia and autism, my brain was half asleep at the time, perhaps busy daydreaming, suddenly lighten up like the sun at noon time. Daydreaming can be a very sweet thing, especially when boys dream with a girl that exists in dreams only. Excessive daydreaming can be an escape from reality and challenges of everyday life situations. But poets, story tellers and novel writers claim their best literary work was done during daydreaming.

The brain, just 2 percent of your body mass, but...

The brain, the daydreaming part of your body, is only two percent of your body mass (not always true if you are an obese person); but that brain of yours consumes twenty percent of the calories you and I eat. I could not help it, but my mind jumped immediately to implicate the replication, transcription and translation processes RNA is involved with to form proteins in my body. The mitochondria and its energy producing molecule, ATP, popped up in my thinking processes and alerted me not to count them out. The truth is that the resting brain, as it is during sleep, especially, REM sleep, is not resting at all. REM stands for rapid eye movement. During my REM sleep, in all probability, my brain is consuming more energy than while I write these lines. I repeat over again, the brain with only 2 percent of the body mass, when it comes to calories consumption, it seems more of a vacuum cleaner that gobbles up 1/5 of all converted calories you have sent down your throat. Even more frightening yet, your brain does not store energy for leaner times. If calories are needed and there is no food for you to bite or swallow a drink and throw it down your digestive pipe, fat that has been stored in your belly, buttocks, thighs, among others, will soon be going to feed your brain. If for some reason you contemplate and risk going on strike against the brain demand for calories, you are on the losing side of the battle. Your brain has a mechanism that will convert what you have into calories until there is no more to devour and death arrives.

The scenario we presented you with a brain need for a huge amount of calories is not a hypothetical case. A group of researchers whose curiosity in that brain of yours need for calories, wanted to know why and how does your brain consume so much energy even during the resting state of brain functioning. To tackle this curious issue they have used the most modern technology at their disposal, PET and MRI scanners. During one experiment, brain cells were fully activated during rest time, but quiet- seemingly resting- as soon as the person began to mentally work on an exercise. Seemingly paradoxically, but a very interesting experiment, it was. During this experiment; it seems that if I work on puzzle solving, my brain cells leave me alone and go on its own business; but if I go into rest stage, the brain neurons go to work.

Raichle, Shulman and the default network mode

Gordon Shulman, a researcher, went through 134 scans and found that regardless whether the subjects in the experiment were reading or watching shapes on a screen, the same group of brain cells began to darken out-- were getting dim-- as soon as they began mental concentration. The researcher called it the default mode. Does it mean that brain cells chat among themselves when not busy doing routine day life chorus and complex intellectual work? This is even more curious. During the resting or chatting time, parts of the network consumed 30 percent more calories, gram for gram, than nearly any other part of the brain. [46] Well, you can argue that the brain multiple neuronal processes including electro-genesis, memory formation and storage as well as energy supervising autonomous nervous system demands caloric

---

[46] - New Scientist, Nov. 8-14, 2008, p. 29.

energy in vast amounts. Our brain is always busy in mind traveling, solving routine daily life chores, planning self-protection or trying to understand Einstein quantum theory. That brain of yours deserves a lot of respect; it led you out of the jungle and now is planning to take you out of planet earth and go where no man has gone before. These researchers used very modern tools to come to very interesting comments and conclusion we cannot avoid sharing it with you. Marcus Raichle says that there is a huge amount of activity in the resting brain unaccounted for. Making use of PET scanning, he noticed some brain areas seemed to get full tilt during rest, but quietened down as soon as the person exercised. Neurons chattered non-stop to one another when an individual is unoccupied, but as soon as a task requiring focused attention, chatting stops or quiets down. We became even more interested in this discovery when we learned that the median prefrontal cortex and the hippocampus become involved in this default network. You already know that the hippocampus is involved in the formation and retrieval of memories. The prefrontal cortex is highly engaged in decision making after receiving input from many important brain areas. A possible task for the default network would be linking multiple neuronal circuits and pathways among clusters of brain cells and provide the brain with an inner rehearsal for considering future actions and choices. Using f MRI, Malia Mason of Dartmouth College in New Hampshire equated this network mode with daydreaming. Randy Buckner and Daniel Gilbert from Harvard University "see it as the ultimate tool for incorporating lessons learned in the past into our plans for the future." Marcus Raichle now believes that the default network is involved, selectively storing and updating memories based on their importance from a personal perspective. He added that rather than burning the extra glucose for energy, it uses as a raw material for making the amino acids and neurotransmitter it needs to build and maintain synapses. [47]

[47] - New Scientist, Nov. 8- 14, 2006, p. 31.

Dr. Marcus Raichle opinion

However, another scientist using MRI reported that daydreaming took place when the default network was active. Default mode was coined for an unoccupied brain. The brain is not working on your own business. It must mean that you are consciously working on something. Marcus Raichle from Washington University at St. Louis believes that the default network is involved selectively storing and updating memories. Another discovery of Reichle on our brain is that it has a sweet tooth; it devours huge amount of glucose, way out of proportion of oxygen it uses. He believes that the extra glucose the brain consumes is use as raw material for the making of amino acids and neurotransmitters that the brain needs to function properly. Now comes the big surprise, Raichle and two more researchers found that the default network's pattern activity is disrupted in patients with Alzheimer brain disease. The author adds that most Alzheimer patient's brain are not occupied with anything because their memories are gone and their neurons are blocked by plaques, and the organelles inside the neurons are entangled one another, and each one unable to send or receive messages.

The point that I am making is that chatting among brain cells in people suffering from advanced Alzheimer disease is very questionable. However, Raichle findings indicated that the default network also turns out to be disrupted in other brain maladies including depression, attention deficit hyperactive disorder, autism and schizophrenia. [48] Neuro-scientist Elkhonon Goldberg, from New York University commenting on the frontal lobe and Alzheimer wrote, My collegians and I have shown that frontal lobes become dysfunctional at a very early stage of Alzheimer's type dementia...The frontal lobes are more vulnerable and are affected in a broader range of brain disorders: neuro-developmental, neuro-psychiatric, and so on,

---

[48] -New Scientist, 11/08/ 2008, p. 31

than any other part of the brain."[49] The findings in this article made me scratch my left ear because in schizophrenic patients, A.D.H.D, and autism in a lesser degree, are on high gear most of the time while in major depression, brain activity as shown in scanners seems to be in siesta time most of the time. The way we see it, brain neurons in schizophrenics and attention deficit hyperactive disorders would be chatting most of times forming hallucinations and delusions and many more thoughts disorders while the deeply depressed person complains that she or he has nothing to think about to report during psycho-therapeutic sessions. Many depressed patients complain that his or her brain is empty. I still recall them saying to me, my head is like an empty shell, nothing goes in and nothing comes out.

Pete, my autistic boy in my office

Our experience with autistic children has been limited, but we observed that they enjoyed themselves when we gave them a game or puzzle for him to work on and, I just sat quietly next to him. We remember a young boy who used to tell me the kind of puzzle he used to love to work on. He began trusting me taking little things home from my office and returned it the following session. He strongly disliked noise and excessive sunlight. He used to pull down the shades in my office. Another interesting observation I remember from his behavior is that he did not like his brother or his mother to share our sessions. During most of the time we spent together; he began our conversation and, at the end of the session, he often asked me for an object to take home. It took me time to refer him to a a licensed trained therapist. I had neither the training nor the neurological understanding underlying autistic behavior. We can interpret him taking home toys from my office as taking home part of me, thus bending isolation through trust. How to use brain plasticity to normalize brain circuits and chemistry is

---

[49] -The Executive Brain, Elkhonon, Goldberg, Oxford University Press, 2001, p. 115

the key question that I was always trying to deal with while in practice.

Autistic children are hypersensitive to sound and touch. For them, a soothing bird song, sounds like a drill sergeant during basic training, or an 18 wheeler truck blowing its horn while I drive my four piston small car in the highway. Most recent scan studies have revealed that autistic children process sound in an abnormal fashion. Likewise, a soft skin touch to an autistic child seems to be the rough game of soccer or football when they jump one on top of the other. "Autism is largely an inherited condition. If one identical twin is autistic, there is an 80 to 90 percent chance the other twin will be autistic as well." [50] Autistic children do not seem to be able to discriminate, consequently, tolerate, degrees of stimuli. It seems to us that the autistic brain closed up neuronal circuit and pathways in selective brain areas ahead of time before it could establish stimuli control levels. They are unable to decrease the sound force of a thunder in the sky and bring it down to tolerance level. Sound frequencies have to be established. Special training and teaching in general will help to re-educate neuronal circuits and pathways leading to normal responses. Autism is much more prevalent in boys than girls. The autistic child is not avoiding or rejecting anyone; he or she is protecting himself or herself from experienced harmful stimuli. The earliest we establish the diagnosis, the sooner we can begin synaptic re-education.

Isolation and loneliness in Schizos and autistic individuals

Schizophrenic and autistic individuals are socially isolated people. Their brain seems to be on the alert tone most of the time. It seems to be as if they are expecting rejection and escaping into their own shell which is comforting and safe.

---

[50] -The Brain that changes itself, Norman Doidge, Penguin Books, N.Y. 2007, p.77.

Researchers have found that the amygdala as well as the prefrontal cortex are more active than most of the rest of brain regions. We may also interpret the hyperactive amygdala and prefrontal cortex as an adaptation or adjustment of those areas to the behavior towards them by people they come in contact with. There may be a biological predisposition that is triggered or activated by outside behavior. Their brain is not engage in regular daily problems and conversation, consequently, they turn inward. We cannot call them daydreamers because the brain is very active forming or recalling hallucinations and delusions. You will find similar brain behavior in bipolar-manic phase individuals. The prefrontal cortex is active judging/ rejection signals, therefore, common sense behavior, decision making and academic accomplishment become deficient not because they are mentally retarded; but because the brain is on the super- alert switch. This may lead to feeling of chronic sadness; not that sadness is inherited, but along with loneliness, is an undesirable behavior coming from a biological predisposition. The environment, particularly, during the early years of development is very critical. We must remember that the brain is not a rigidly wired organ of our body. We have new neurons forming every day; particularly, in the hippocampus and ventricles. In addition, we have synapses connecting neurons and establishing new pathways reinforcing good and appropriate behavior and thinking. Brain plasticity is for us to take advantage of and put to work with people in need. The brain learns new things every day; it learns something new and the region or cluster of neurons responsible for processing sound and touch in autistic children have to adapt to new levels of signals.

Neurobiologist Jeff Lichtman from Harvard University hopes that studying the structural changes in the brain will eventually show him how information is encoded in our wiring. Using a method he dubbed Brainbow, a small piece of DNA with genes that code for random amounts of yellow, blue and red fluorescent proteins in nerve cells are inserted into mice.

Through this method or research tool, his team can give individual neurons different colors; the colors will allow researchers the opportunity to determine which synapse corresponds to which axons. Connections offer a way to investigate thought disorders, such as mental illness and learning disabilities. Further addressing the wiring or connections between neurons, he says: The plasticity of the brain's wiring is one of its greatest strength and scientists hoping to learn more about how we learn and adapt as we grow."[51]

Plasticity of the brain

Aristotle, the ancient Greek philosopher and tutor to Alexander the Great, proclaimed some truth on personal observation that became absolute dogmas for almost two thousand years. The Sun and stars were moving around planet earth as if it were the center of the universe. Our dear and sweet mother earth was believed to be a flat mass of land until a few hundred years ago when Christopher Columbus helped tear down that fallacy. People become dogmatic and stick to beliefs and practice even when facts contradict it. And, we should point out that it happens not only to the uneducated masses. The theory of localization as pronounced by Paul Broca and Carl Wernicke became an axiom, an unchallenged truth still persisting in some professional quarters. Dr. Broca had a patient that answered all his questions with the word, Tan, tan after suffering from a stroke in 1861. After the patient's death, he performed surgery and found out that an area in the left hemisphere now known as Broca area was gone. A few years later, Dr. Carl Wernicke had a patient with similar problem, except that this time, Wernicke patient spoke in an unintelligible fashion; it sounded more like gibberish than a normal conversation. It lacked conjunctions among words and phrases. After surgery, a region in the left upper temporal lobe

---

[51] - Science Illustrated, March-April, 2010, p.35.

121

now known as Wernicke area was caput; it was damaged. With this two cases followed by many well attended professional conferences, the one function, one location theory was established. Both areas: Broca and Wernicke are located in the left hemisphere of the brain, thus, another "truth" was established, and the left hemisphere is dominant for language.

A non-dogma believer, J. Cotard

However, a non-dogma believer and science oriented physician, Jules Cotard, in 1868, studied children who had early massive disease, in which the left hemisphere, including Boca's area, wasted away. Yet, these children could speak normally. In 1876, Otto Soltman removed the motor cortex from infant dogs and rabbits. The motor cortex is responsible for movement. We will address this shortly. Another surprise, he found that his little rabbits and dogs could move. Both, Cotard and Soltman had not only disproved the dogma one function, one location, but more importantly, they had discovered the plasticity of the brain. Both experiments had shown that the brain was plastic enough to re-organize itself to meet demands of daily life chores. From Broca's surgery in 1861 until now have passed 150 years, and we still have many professionals doubting the brain capacity to take over functions damaged by strokes or a disease. Furthermore, despite massive evidence using most modern scanners, it is very hard for some individual to accept the growth of new neurons in the brain. Plasticity of the brain must be understood not only taking place at synaptic level, but clusters of neurons taking over functions from one brain hemisphere to another.

The year 1876 in medicine is a long time ago.

In reference to Otto Stoltman work with kittens and puppies motor cortex, we do not know what did he do? It is a complex system composed of multiple regions in the brain. His

122

surgery was performed on infant animals when their brains wiring was taking place or in the process of consolidating functions. What areas or groups of motor neurons he removed is unknown to us. It was 1876 and none of the machines we now have to slice and examine brain organs, tissues and functions existed at the time. We decided to write a few observations on this subject so you can appreciate the strain efforts a Parkinson affected person goes through to be able to get up, move and walk. Similarly, we have family members, friends and valiant soldiers that come home with spinal and brain injuries. The motor cortex has multiple loops facilitating lateral horizontal and ascending and descending connections with the neo-cortex and spinal cord, among others. It is divided into three reciprocally interconnected areas, the primary cortex, supplementary motor area and pre-motor area. In the primary motor area, neurons correlate with a variety of movements and firing of individual cells. The secondary motor cortex is involved in planning movements and the pre-motor area is particularly concerned with planning movements that require sensory cues. [52] With Parkinson patients and combat soldiers with central nervous system injuries; it is very painful to overcome the limitations imposed by this brain deficits. It is not only the physical, but the psychological stress coupled with the injury. These are adults that have to learn anew how to get up, move and walk. They have to force neurons in different brain locations to learn to do a job they had not done before. It is a physiological and psychological demand on the patient to force plasticity of brain cells to take a new function. It is a feeling I cannot share with you in words when I see patients, family members and friends walking and doing things they could not do before.

[52] - Neuroscience, A. Longstaff, Bios Publications, U.K, 2000, p.221-22.

P. Bach –y- Rita in the journal Nature.

In 1969, Europe most prestigious scientific journal, Nature published an article that made many people shake their heads in disbelief. The article, whose leading author was scientist and physician, Paul Bach-y-Rita described a device that enabled people who had been blind from birth to see. My initial response would be that he was talking about Sy Fy or the biblical story of Jesus of Nazareth two thousand years ago; however, it has been proven that it was neither. Almost half a century ago, Doctor Rita had rejected the above dogma and proclaimed that our senses have an unexpectedly plastic nature, and if one is damaged, another can sometimes take over for it, a process he called sensory substitution. He says, we see with our brain, not with our eyes. It reminds me of Jody in Captain Piccard starship, Enterprise. Dr. Rita developed ways of triggering sensory substitution and devices that give us super senses. He laid the groundwork for the greatest hope for the blind: retinal implants, which can be surgical inserted into the eye.[53]

Self-isolation, not loneliness

Self-isolation was widely practiced by monks during early Christianity. Socrates dual nature of humanity, body and soul, was well accepted by Christian to restrain bodily imperfection and desires and advance the soul journey while on planet earth. Today, we advance the needs of the body and the thinking processes of our brain. Self-imposed isolation, except during hibernation in some animals is not our preferred way of life. However, our curiosity for all living things stopped at a bug buried in ice for 120, 000 years; yes, one hundred and twenty thousand years. Scientist coaxed it back to life. The bug was dormant in three kilometers deep in the Greenland ice

---

[53] -The Brain that Changes itself, Norman Doidge, Penguin Books, N.Y, 2007. pgs. 10-17.

sheet. We asked ourselves, what did it eat to survive so many years under such an inhospitable environment? The research team leader, Jennifer L. Curtze named dust, bacterial cells, fungal spores, minerals and other organic debris as source of food. She further added that the best medium to preserve amino acids, organic compounds and cells is ice.[54] We may add that Jupiter moon, Europa, and poles of Mars may well be home for primitive life forms to thrive. In 2010, scientists from NASA reported that a bacterium in a lake somewhere in California could survive in arsenic acid which is fatal for all of us. This is not SY Fy or a matter of faith; these are scientific facts. The journal, New Scientist dated 10/18/2008, on page 17 reported: DNA of gold- mine bug may be key to alien life. The article claims that "the organism ability to live in complete isolation from other species, or even light or oxygen, suggest it could be the key to life on other planets." This strange bug was found in a South Africa gold mine nearly three kilometers beneath earth's surface. Dylan Chivian from Lawrence Berkeley National Laboratory in California analyzed the newly found bug and said that the bacterium gets its energy from radioactive decay of uranium in the surrounding rocks; it has genes to extract carbon and nitrogen from the environment-both essential for protein synthesis."

Evolution, adaptation and surprises of all sorts continue to keep my brain in a permanent state of alert for new discoveries. In an October 2007 edition of New Scientist, I read that there is a possum that slept for a year. The researcher claims that possum he was experimenting with, after stuffing itself with food, curled up and went to sleep for a full 367 days in his laboratory. Another animal named, Zapus princeps, had slept for 320 days, however, an Australian pygmy possum broke record by using one-fortieth of the energy it does while awake.

---

[54] -New Scientist) 6/20/ 2009, p. 8.

The hunt for diseases and the genome sequencing

The hunt for diseases genes has engaged many outstanding scientists around the world. Initially, the main focus of attention was centralized in DNA letters changes or mutations that were suspected of fatal diseases such as multiple sclerosis, Alzheimer dementia, schizophrenia and autism, among others. In some of these cases, the gene may appear normal, but there might have been a surplus or deficit of DNA sequences. Among copy number variations we came across chromosome 21 with three copies instead of the normal pair, consequently, provoking Down syndrome. Past literature refers to this syndrome as Mongolian looking children. This type of rare variation was considered conducive to diseases in most case, however, further studies revealed that variations in genes quantity is not an anomaly, but on the contrary, is quite common. In the year 2006, a group of geneticist analyzed DNA from 270 individuals and identified an average of 47 copy number variants per person. The genome sequence of C. Venter, a pioneer in the field, was found to have 62 copy number variants. Already mentioned in another chapter, J.D. Watson genome also revealed a good number of copy number variants. These two outstanding scientists are doing exceptionally good work for humanity; so you cannot, but wonder what is the role or function of those rare variants. However, "copy number variants have been linked to disease like autism and schizophrenia."[55] We can argue that Down syndrome rare variants should not be placed along autism and schizophrenia, and we fully agree with it. We have made the observation several times; this last two diseases as well as bipolar we consider it syndromes. It is not only a matter of genes, proteins, cells signaling, neurotransmitters and glia cells, but a myriad of other causal factors. By the way, copy number variations have been described as alterations of the DNA of a genome that provokes the cell to have an abnormal number of

[55] - Scientific American, 06/ 2009, p. 24-25.

copies of one or more segments of the DNA molecule. The alteration may be an increase or deletion of a segment of the genome. In some cases, the variations may each be as much as 10 percent of the human genome posing possible serious risk factors for diseases or abnormalities nobody wants to have. The scientific opinion on the etiology of these genome variations seems to be divided at present time. There is evidence that it is inherited while there is scientific support for de novo mutation, meaning that it may be seen as a spontaneous mutation.

Mobility within the genome

We must have clear in our brains that the genome is not a long strand of beads tied to each other without mobility. The genome suffers deletions, modifications as in SNPs, duplications as well as multiple interventions by genome fragments. There are several gene segments that are involved in DNA-RNA multiple processes, consequently, variations in general seem to be the rule instead of the exemption. Copy number may take place in a single gene or be extended to a neighboring set of genes. The quantity factor, meaning too many or too few in relation to risk propensity genes may be responsible for a good substantial amount of human phenotype variability, behavioral traits and disease provoking sensitivity.

James Lupski, a clinical geneticist at Baylor College of Medicine in Houston, Texas published online (10/14/2010) in the prestigious scientific journal, Nature, "highlighting the importance of rare genetic variants in causing diseases...even one copy of a certain gene can have profound consequences for brain development and mental disabilities." We add, How about gene mutations that do not code for proteins and become degenerated debris in a cell? The mitochondria DNA also is subject to multiple modifications and problems. This female contribution to our body and heritage can be a blessing or a liability. Dominant and recessive variants on both sides can contribute to make me a genius or a disabled individual unable to care for myself. One of my parents gave me an unwanted present, thalassemia minor.

Reinforcing erroneous assumptions

Not long ago, many scientists used to believe that silent mutations were inconsequential to health because those changes in DNA would not interfere with protein synthesis. Although some disorders were traced back to a silent mutation, the general consensus was that the culprit could not be a silent mutation. It was argued that it would go against the general observation of most researchers in the area of molecular biology. Reinforcing that erroneous assumption was the discovery that many silent mutations in various species were preserved over a long period of time. In addition, in many species the consequential change made protein more efficiently. However, there was a pronounced and significant exception; it did not apply or did not work in humans. As the human genome sequencing technology advanced, scientists began to question their erroneous assumption, an assumption based on scientific observation and logical conclusion. By now, you are fully aware and most likely convinced of the efficient processes of protein synthesis. You recall that in RNA chains, the letter U for uracil is substituted; meaning changed for the letter T thymine in DNA chains. During this complex process of protein manufacturing there is a letter or nucleotide change for the good or benefit of life in general. Simply reasoning it out, the information encoded in nucleotides-those blessing letters-is converted into the language that builds or forms amino acids which in turn will finish up as a protein. During the whole process of manufacturing proteins there is a set of rules and steps that should not be broken. This set of rules governing DNA-RNA is widely known as the genetic code. Among the rules-code- is the formation of messenger RNA, and its editing process(splicing non-coding DNA), transfer RNA that bears and carries amino acids to its destination, the ribosomes. The job is not as easy as many people think it is. Take for instance the amino acid leucine. It has six codons, all coding for the same amino acid. The codons are: UUA, UUG, CUA, CUC, CUG, and CUU. On the other hand, you have the amino acid tryptophan which has only one codon, namely, UGG. You have the amino acid serine with six codons, and tyrosine with two codons. In all, there are 64 codons codifying for the twenty amino acids. As I said earlier, this set of rules cannot be broken or altered; otherwise we will end up with multiple problems and diseases. Do not be frightened, we already told you that the cell possess a self-correcting system to ameliorate the problem. Just a few more observations on this subject may be helpful to some students that like to go a little further on gene mutations and related issues in biology. A single letter change to the code or rules we have talked about that ends up as a healthy-efficient protein, known among neuro-scientists as point mutation, can provoke a changed codon forming a wrong amino acid, which in professional language is known as non-sense mutation. Consequently, the resulting product would be a shortened protein. Therefore, a single letter change in the chains we are writing about it; can cause several problems including a stop codon which will provoke a change in amino acid culminating in a longer or different protein. No wonder scientists in this field of molecular biology could not point with certainty the locus of the problems and come out with the correct conclusion.

In retrospect, scientists had begun to challenge the prevailing theory on silent mutations about thirty years ago. By the middle of the 1980s, scientists began to realize and accept that silent mutation could intervene and improve protein synthesis, at least in single cell organisms. With new and improved research technology, scientists began to discover that cells had preference for some codons, meaning that cells did not treat codons on equal basis. It seems that cells were employing protein synthesis efficiency selectively. Just as we tend to care for and breed animals and birds that satisfies our curiosity, needs and interest, relatively recent research found out and came to a rational conclusion that mammalian genes tend to favor certain codons.

To please my tormentor who do not stop asking me to repeat things for him, I will say that the genetic code that we have referred to many times, is nothing more than a set of rules or code in a cell that dictates or govern how information encoded in genetic material such as DNA or RNA sequences will turn out as amino acids, and ultimately, as proteins out from ribosomes. The in- between processes of nucleotides, codons or triplets, transcriptions and translations making a stop at ribosomes are indispensable rules that the gene code has imposed in the process before the final product, your body building blocks: proteins that will be delivered to specific stations-organs- of your body. You owe to yourself to learn how you are constantly re-building yourself to keep your body beautiful, handsome, healthy and intelligent.

A BUG INSIDE MY BRAIN?

Just a few words to get my brain-mind- clear of something that is bugging me for a while now. During our essay it seems we have picked on the translation process as the bad boy or villain of the gang responsible for many of our diseases and disorders. Believe me please, I am not prejudiced against translators or anyone else. On the contrary, we recognize how difficult it is to get an abstract concept and decode it into understanding everyday parlance. Sequencing the human genome and learning about all the debris, markers, mutations, modifications, intervention by short segment of genes, etc. makes me admire their job. Discovering and translating for us short nucleotide sequences responsible for flagging or alerting boundaries of the exon to cellular editing instructional machinery is awesome and worthy of praise. This editing process becomes even more interesting while you find

SNPs (short nucleotide poly-phormisms) engaged in or influencing this already complex process of protein manufacture. During this scenario there is no room to blame translators or researchers; it is the gene code itself as expressed in DNA and RNA. We have said several times that brain matter inside your skull is basically composed of glia cells. Some experts in the field claim that over 90 percent of your brain matter is not involved in making good and useful decisions for you. Few scientists and writers that like to speculate, claim that around 93 percent of our bodily flesh, but in particular, our brain is a left over from evolution. The human brain can be seen as a huge mass of cells, glia cells, occupying much needed space and consuming a high proportion of sugar and oxygen neurons could use for the benefit of our body. A one 100 billion neurons are busy working for you 24/7 while around 850 billion glia cells seem to be taking a long nap. However, some glia cells out of that huge number are part of our brain blood barrier; they are performing as inspectors of blood content delivery to all cells in the brain. There are other equally important jobs glia cells are responsible for, but as far as I am concern; it is the comparable small number of neurons that keep me busy learning about myself. Bugging my brain-mind-is consciousness and unconsciousness. In my book, An Episodic Toxic Memory, a survival story, I attempted to explore both subjects, but unfortunately, it became short of a satisfactory answer. We hope neuro-scientists can tackle the issue and provide a satisfactory explanation for most of us. You may suggest that we should begin with single cell organisms, which we have, while Peter at the end of the table may rightfully argue that single cell organism do not have the complexity of a human brain. John, a faithful believer in a supreme being, would intervene saying science could never decipher the mystery of the brain because man's soul cannot be placed on a petri- dish, test tube or seen through a PET or fMRI scanners. We have attended several meetings on the above subjects. We have agreed to have more meetings, but opinions remained as far apart as we began the first day.

You are unique in our solar system

We have no doubt that we are the most beautiful and intelligent creature inhabiting our solar system. Cells making your body took many years to develop and build the most complicated and beautiful organism existing in our solar system.

We are an intelligent species with vast knowledge about ourselves and everything existing on planet earth. We know a lot about our sister planets and sustaining star, our sun. Our astrophysicists have calculated that we have a few more billion years to live before it will change into something else. We have developed sophisticated tools to examine and study ourselves including the most complex organ in our galaxy, our brain. This is a little far fetch extending our imagination, but at present time no one has challenged it. Just a decade ago we completed the sequencing of the human genome, something that not even gods had attempted to do it. We engaged in a research scientific race to find deficient genes- coding and non-coding genes-with its variants, DNA markers that are linked to diseases. In the past, we pleaded gods, shamans, healers of all kinds and even religious icons. Today we have invented first rate laboratories and sophisticated machines to probe our atoms, molecules and cells of our body. We have made progress, but much remains to learn about the intricacy of cell life in our organism. Naive promises were made to the public, especially to individuals and families with chronic degenerative diseases hoping for a miracle pill or new form of scientifically proven drug that would end their misery. Locating the gene or genes responsible for the etiology of our diseases has been quite elusive. However, we have made significant progress in all areas of medical technology and science in general if we look back fifty or sixty years ago.

Committed and enthusiastic, but naive

Rather naively, but fully committed and enthusiastic, we were hoping that clear cut diseases like Parkinson, Alzheimer, multiple sclerosis, and to lesser extent, even schizophrenia, bipolar an cancer would be eradicated from planet earth. During the 1990s many prediction were made by lay people as well as some scientists; however, it proved to be poor judgment and unproven science. Hardly any substantive treatment progress has been made or found in the above mentioned human maladies. Further hampering our disease hunt scientific race, funding was cut in many areas of research and, concern citizens negatively influenced congressional legislators over stem cells research monies, among others. Besides premature expectations and limited funding, we believe the scientific community working in the sequencing of the human genome did not expect to find that most of our genome is composed of non-coding genes and the problems it presented to them. Even after discovering the so call junk DNA, we did not know what to make of it. A few theories have been advanced, but the interrogation remains largely unanswered. Similarly, DNA markers, epigenetics and much more, have slowed down our race in the hunt for the origin of human diseases. There is a mountain of specialized research that needs to be translated into therapeutic tools and drugs. We applaud Francis Collins from the NIH for taking wise steps in this direction. Mel Graves, a cell biologist writing for Nature said: "We fooled ourselves into thinking the genome was going to be transparent, blue print, but it is not. Instead, sequencing and other new technology spew forth data, the complexity of biology has seemed to grow by order of magnitude. [56]

In the same article on Nature (-4/01/2010, # 7289), Eric C. Hayden wrote, "Much of the non-coding DNA has a regulatory role, small RNA of different varieties seem to control gene expression at the level of both DNA and RNA

---

[56] -Nature, Vol.464, # 7289, 04/01/2010, p 664.

transcripts in ways that are still only beginning to become clear. (He used the words seem and beginning to verbalize his opinion). The author of the article, E.C. Hayden quoted another scientist, a mathematical biologist who said: "Just the sheer existence of these exotic regulators suggest that our understanding about the most basic things- such as how a cell turns on and off- is incredibly naive." In contrast, some of our television cartoons make it easy for us to continue to entertain unrealistic promises. A TV song goes like this, Triplets nucleotides UGG codes for tryptophan; it loves its neighbor AUG methionine with three letters each. (Repeat part of the above) The song continues: Not far away we find leucine that can come out in six different triplets. All these letters are translated into amino acids that will become proteins for my body. I found the song very interesting and educational. But, I could not help stop thinking about how far removed from the reality of the process it is. Protein synthesis does not come as easy as presented in the classical DNA-RNA-protein dogma of our double helix discoverers icons, Watson and Crick.

Protein encoding problems

The song goes on naming more nucleotides and amino acids. She deserves our admiration and recognition for introducing the code of life and, thus biology and genetics, at an early phase of life. I wish I had a science minded teacher during my early school years. However, the reality as we have found out at present time is not that simple. The genetic instructions of our complex organism have proven to be more difficult to decipher than we had anticipated. Nucleotides encoding for our proteins encountered a myriad of problems still largely unknown. Take, for instance- introns- the non-coding portion of our genome; they have to be identified, cut and removed from the mRNA precursor before it can be considered for protein synthesis. Both, the non-coding genes and the splicing processes are not fully understood to be able to

134

relate to disease genesis. Furthermore, it is easy to talk about replication, transcription and translation processes during protein manufacturing; but connecting it to a syndrome or clear cut disease, is basically an unrealized dream in the horizon of medical research. Nonetheless, DNA technology has become a helpful ally of medicine.

For us, removing introns from exons in pre-mRNA is seen as a wonderful trick of our cells that needs further research in the protein manufacturing process. If we consider pre-mRNA first step and all the steps in between until we get mRNA and tRNA transported to a ribosome culminating in very specialized protein in the prefrontal cortex; it is a very long, long trip. The same thing can be said about a specialized protein in major organs of our body. Most researchers have voiced their concern identifying a loci or a simple step during protein synthesis that would end up in a deficient organ or tissue without considering many other causal factors in disease genesis. Our laboratory companion, the mouse, is extremely helpful and relatively easily engineered to carry out our experiments, but a schizophrenic or bipolar mouse, despite its almost similar genome, is not a human being growing up with early symptoms that later will be a full blown schizophrenic. The intricacy of a gene generating a multitude of messages has us wondering about a simple presentation of our genome and undelivered past promises. We are working very hard, but the task ahead is a challenge in need of bright brains and commitment. We have been focusing our attention on rare variations that applies to population worldwide. How about relatively new mutations, modifications and adjustments maneuvers that boost up human evolution, but might still be in the process of making proper and adequate accommodations? Our hippocampus and prefrontal cortex are composed of groups of neurons under constant, persistent and damaging stressful conditions. Residents of urban areas, especially, after the industrial revolution, are highly exposed to stressors not even imagined by our own great grand-parents. Our entire body, but in particular, our brain, is making relevant adjustment to multiple demands all times to be able to survive. How exons and introns interact in critical regions of our brain and bodily cells to come out with appropriate signals for an appropriate, timely and efficient response, are questions that keep on bugging our brain. Introns and glia cells are demanding more attention from all researchers. In addition, we are not only referring to a specific locus in a cell or organ. We are contemplating the influence played by an incomplete or wrong signal as it relates to healthy and non-so-healthy genes. We have in mind syndromes like schizophrenia, autism and bipolar depression. We see it not as a cut and dry disease, but a mountain of symptoms contributing to a very painful brain disorder.

David Lewis and schizophrenia roots

Leading researcher, David Lewis has spent the past two decades exploring schizophrenia's developmental roots. He runs one of the more comprehensive and sustained attempts to explore normal and pre-schizophrenic adolescent brains. Dr. Mayberg at Emory University described it as, "One of the smartest, most creative and most promising angles I know of on schizophrenia." Dr. Lewis focused his attention on the prefrontal cortex we mentioned many times in research studies. On above paragraphs and chapters, we have told you how this region of the brain is connected and receives a myriad of projections from the limbic system including the amygdala, hippocampus as well as sub-cortical and neo-cortical input. It is often called the CEO of the brain, but it seems partially true because it is dependent on emotions coming from the limbic system to make wise decision. Furthermore, you have the life saver amygdala that has a mind of its own when it comes to protect you from immediate danger. Dr. Lewis work examines the relationship between two types of cells: pyramidal neurons and chandelier cells. He thinks that chandelier cells fail to cultivate pyramidal cells during childhood or early adolescence, and the communication needed in the region for a robust connection, leaves the prefrontal cortex incapable of coordinated firing and synchronicity. Tom Insel from the NIH says Lewis model provides something this field really needed, a framework for linking observation at the molecular, cellular and system levels. [57] Coordination and cooperation among researchers, is paramount to be able to tackle syndrome like diseases. With brain diseases and disorders, we hope researchers do not fail to include the functions of the majority of its cells, glia. Astrocytes seem to be eavesdropping on neurons at all time. Oligondendrocytes are responsible for myelin on neuron's axons, among other roles.

Atoms and molecules interaction

When we observe the multiple organelles within a cell, we wonder how atoms and molecules interact with each other to function normally. Molecules are constantly bouncing against each other in an aqueous environment in animal cells. The same must hold true for atoms, but this elusive and microscopic particle of every element is very hard for us to advance an intelligent opinion. Atoms and sub-atomic particles are constantly traveling at huge speed through, above and under me. You are not exempt or free from this constant bombardment of atoms and subatomic particles either. Let physicists do the explaining for us. However, collision among molecules and atoms of all sorts within and outside our cells produce force, meaning energy. You and me may wonder if the abundance of force floating around molecules of our basic elements like hydrogen, carbon, oxygen and nitrogen are not overly exposed to degeneration or degradation into their basic atomic components and be responsible for some of our diseases. This intelligent observation is an interesting one and deserves serious consideration by concern researchers. Of

[57] -Nature, Vol. 468, 11/11/2010, p. 155-56.

course, there are multiple chemical reactions going on inside our cells necessary for normal function. Our body is in a continuous process of self-renewal (apoptosis and cell division and multiplication), but we are referring beyond and above the normal birth to the aging phase of our life. Or, is it naive to separate one from another? We are thinking in line with our basic molecules forming covalent bonds that are many times stronger than floating molecules or atoms. Our macro-molecule DNA with the assistance of another equally powerful molecule is of paramount importance to you and me as RNA produces proteins called enzymes. These enzymes are responsible for the making and breaking down covalent bonds. In other words, tearing down the union of those life indispensable units of our organism formed to tolerate a lot of tension and stress under critical conditions; can come under attack by a protein-enzyme coming from an incomplete or dysfunctional gene. This is a very valuable truth to remember while keeping an eye on our health. There are many genes in our genome called introns, that we do not know its function or functions. There are more non-coding genes than protein producing genes. You already know that introns are clean out before messenger RNA is ready to be shipped out to become part of a protein. However, you also know the multiple problems and failures a protein encounters before its final step is completed.

Covalent bonds

If you allow us to go back for a little while to try to make our point clear to you, we will appreciate it. Let's begin by saying that covalent bonds are made when atoms share electrons. You may recall your elementary school teacher telling you that atoms are composed of protons with a positive charge and neutrons which are electrically neutral, and are located in the nucleus. The teacher also taught you and me that atoms have electrons traveling around the nucleus in an orbital fashion. Electrons carry a negative electrical charge. Generally, atoms have two types of sub-atomic particles in the nucleus, namely, protons and neutrons with an equal amount of protons and electrons. Hydrogen has a positive charge proton with an electron, but no neutron. The most abundant atom in our body is the hydrogen atom; the lightest element that became part of our organism. A basic rule or law of physics to remember is that the electrical charge possessed by each proton is exactly equal, but opposite by that carried or possessed by a single electron. This positive/ negative charge of protons and electrons would make the atom electrically neutral. The mass of each neutron in the nucleus is equal to each proton.

Another abundant atom in our body is carbon atoms. In the nucleus, a carbon atom carries six protons which are electrically positive charged, six neutrons and six electrons which carry a negative charge. Most of carbon atoms in planet earth are composed of six protons and six neutrons known as carbon 12. Neutrons contribute to the stability of the atom, but the sub-atomic particle that definitely determines the element each atom belongs to, is the proton. There are carbon 13 and carbon 14 isotopes. Carbon 13 has the normal 6 protons, but 7 neutrons. Carbon 14, a very radio-active isotope, has the normal six protons, but 8 neutrons. The six protons in carbon 12 in its normal condition remain unaltered. If for any reason, atom carbon 12 would degenerate to 5 protons or acquired an extra proton, making it 7, that atom is no longer a carbon atom. However, a couple of pages back we talked about covalent bonds. These bonds are very strong and do not break down that easily. There is a significant difference between sharing an electron between two atoms that fulfill mutual specific needs for each other, and releasing or donating electrons to atoms or molecules. Besides strong covalent bonds, there are hydrogen and ionic bonds that we will not address here. While Watson and Crick were working on the structure of the double helix, one of the mind bugging issues was how the bases inside both backbones came together. Nucleotides from both backbones were united by hydrogen bonds, a weak bond. This hydrogen bond would facilitate the opening of the double helix for DNA replication and transcription. We brought this up in an effort to make you aware of the many complex processes our proteins, organs and tissues go through before they become functionally normal in our body as well as yours.

Now-days, some individuals refer to us as made of carbon chips. What it means is that if we were to do the impossible, meaning, take all the water out of our body, approximately, all the molecules in every cell in my body and your body are based on the carbon element. Carbon atoms are extremely versatile, it can easily form covalent bonds, join other carbon atoms forming large chains and rings for complex molecules. When you look at the double helix, you see the inside bases, adenine, thymine, guanine and cytosine in the form of rings. Carbon atoms are involved in DNA methylation and histones modifications, among many other interventions in protein synthesis and cell normal functions. In the same vein, amino acids are a particularly characterized class of molecules that distinguish it from any other in our organism. They all have an amino acid group (NH_2) and a carboxyl acid group linked to a single carbon commonly called the a- carbon. There is a side chain or molecule known as R attached to this a carbon, thus providing the chemical variety to amino acids. We do not shy away for repeating that proteins are manufactured from 20 amino acids that are vital to life on earth. It is not intended to be a joke, but amino acids and proteins are parts of your body in sickness and in health.

Just as amino acids are the subunits of proteins, nucleotides are subunits of DNA and RNA molecules. Basically, a nucleotide is a molecule compound of a nitrogen-containing ring compound linked to a five carbon sugar. The sugar in this nucleotide can be either ribose or deoxyribose; it carries one or more phosphate group. You see it clearly in the double helix as per Watson and Crick classical illustration.

Although the word nucleotide is commonly used to define or refer to the famous letters ATCG; it also refers to the phosphate-sugar backbone segment linked with each base pair on both sides of the double helix. Nucleotides have additional functions we are not equipped to deal with here. In summary, you have the ribose containing molecule RNA and the deoxyribose molecule; the RNA spews out uracil for an exchange of thymine and DNA equally spews out ATCG.

Amino acids are a very particularly characterized class of molecules that distinguish it from any other in our organism. They all have an amino group, (NH_2) and a carboxylic acid group linked to a single carbon atom, commonly called the a-carbon. There is a peptide chain, known as the R chain, attached to this a-carbon atom, thus providing the chemical variety to amino-acids. These amino acids molecules will be involved in the manufacture or synthesis of proteins. And... these proteins are part of your body in sickness or in health. Just as amino acids are the subunits of proteins, nucleotides are the subunits of our two great molecules, DNA and RNA. Basically, a nucleotide is a molecule composed of a nitrogen-containing ring compound linked to a five carbon sugar. The sugar in this nucleotide can be either ribose or deoxyribose; it carries one or more phosphate group. You can appreciate it clearly in the backbone of the double helix with the bases in between.

Although the word nucleotide is commonly used to refer to the bases, it also refers to the phosphate, sugar backbone segment linked with each base pair on both sides of the double helix. Nucleotides have other functions we are not addressing here.

Teachers love to go over and over again their lesson so it becomes long-term memory on your brain. Do not blame us for being repetitive; consider it as an innocent attempt on our part for you to remember us. My teacher used to hand out hand written declarative sentences like: On a weight basis, macromolecules DNA and RNA are by far the most abundant of the carbon- containing molecules in a living cell. To make her point even more clearly to us, she used to add that all cells in her body were made of dear carbon atoms. Carbon molecules are the most distinctive properties of all living things. At the time, I was wondering about my soul, oxygen, nitrogen, hydrogen and other elements that exist in my body, but she did not emphasized it as much as she did carbon atoms. I remember her as a caring and dedicated teacher that accomplished her goal of creating long term memory in my brain. She hammered DNA and RNA molecules on us because she knew, among other things, that both molecules are engaged in the formation and storage of memories besides transmission of hereditary information. And... involved during this storage and transmission of information are our inherited disease propensity risk we have been writing about. It is truly a long journey from our ancestors thousands of years ago to the present; we have survived not only the jungle, but the armies of microbes, plagues and natural disasters of all kinds. In great urban areas, when we ride public buses and subways, we are breathing the same air that is infected by sick people riding with us. A secondary problem coming out of sharing the same air trapped in subways and elevators is that not everybody has had developed the same defenses against bodily intruders like virus and bacteria that would make us sick. Let us say you were born and raised out in the country, not far from the ocean like me. We were used to eat daily fresh fruit and vegetables picked up from our small farm. We ate a multiple number of beans and roots accompanied by home grown chickens, goats and cow meat usually dried and salted at home. My mother, like her father was a locally well-known botanist that kept a medicinal garden at home for family and neighbor's needs. Although tobacco was one of the main sources of income in our home, no one of us learned to smoke cigarettes. It was exported up North. I do not remember going to a medical doctor, except a dentist, until I was nineteen years old and living in a northern urban area riding trolleys and subways.

Notes from my teachers

From my teacher I carry notes remembering me that the DNA polymerase is trusted with the responsibility of manufacturing new DNA strands by using the "old" strand as a template. This self-replication process provides the mechanism for the addition of new double strand of DNA molecules. DNA replication produces two double helices from an original DNA molecule.

Each of the two strands of DNA is used as a template- a mold- for building complementary strand. On top of a blackboard, she had written: Most genes are short segments of DNA encoding a single protein. The DNA in a gene is not used in its totality encoding a single protein. Short stretches of DNA become busy identifying, determining timing, place and size, among other jobs, in protein manufacture. We do not tire repeating that you have a very wise DNA molecule. Your genome is a large and effective information storage vault that deserves much respect and awesome admiration. Another awesome attribute of our DNA double strand and self-replication is that when it separates for replication, the information is passed on to the complementary strand.

DNA replication process

The DNA replication process is initiated by a protein-enzyme- that attaches itself to the DNA and opens up-unzips-and separates apart each strand by forcing loose or unbound-the hydrogen bonds between the bases. There are particular genes segments responsible for signaling where the replication process begins. During the whole process of replication there is a very wise enzyme known to us as DNA polymerase that is busy directing this complex operation. It synthesizes the newly formed DNA strands using one of the old strands as template. This wise DNA polymerase commands the addition of nucleotides to the growing DNA strand. The term old does not refer to time or age, but to one of the original strands that will serve as a mold for a complementary new strand. For this very reason, each strand of DNA molecule becomes a template for a complementary strand. This enables the cell to copy its genes before passing them on to coming generations. This wonderful process which is carried out within the cell must be done extremely fast and accurate. The beauty of this replication process is that it may include many thousands of nucleotide pairs. This replication machine which is basically composed of proteins, manufactures two complete double helices that are identical to the original DNA. This replication process involving thousands upon thousands of nucleotides must do it within a minimum of errors. Take a minute to think about people who live to be almost a hundred years old. Those molecules responsible for replication, transcription and translation without taking a week or two vacations a year deserve to be respected and praise for a job well done.

Your chemistry, DNA

Getting to know your chemistry, meaning your DNA you will learn the rules that govern all forms of life on planet earth; it is the chemical soup that gave rise to an RNA and DNA world that in a relative harmony with another wise molecule, the mitochondria has produced the most intelligent organism in our solar system. Laws of physics, chemistry and biology seem to work in unison and applicable to the rest of the universe. No matter which element of our body you take; it is composed of atoms, and atoms of various elements are flying or traveling in space in the form of clouds-gases-and become attracted to each other forming celestial bodies like stars. And stars, like humans, are in a constant dynamic process of birth and death ever since a big bang took place somewhere around 18 billion years ago. I am no good at figuring out time, but it happened many, many years ago. Some stars go through the process of implosion-explosion becoming what is known as a supernova. It is a type of recycling process in which the old star goes through a cycle of self-consumption to appear anew in atoms and molecules in another space and time. A few billion years ago, our star, the Sun was born, and in another few billion years more, you can rest assured that it will die. You and I will not be here to witness it; but the same atoms, the same energy that exist in both of us at the present time "alive" will forever exist; except if it falls prey to a black hole, according to most modern astrophysicists.

Tiny particles inside the atom

In an atom of an element there are tiny particles and waves responsible for chemical and atomic processes we are beginning to understand and use for our own benefit. Likewise, in our bodily cells we have chromosomes that are structures within the cell nucleus where genes are carried or stored. There are more genes than chromosomes. It is not only a scientific truth, but a logical conclusion. Consider the genes you must carry for the color of your eyes, skin, hair, and even your height, to some extent, is determined by your genes. These genes you get from your parents are filed in your DNA great and wise molecule waiting expression during mating. The genes you inherited from your ancestors do not fade away, fuse or blend in a fertilized egg. They remain unaltered, or as Gregor Mendel called it, particulate, meaning that those very particular genes are not lost after fertilization. This wise and humble dedicated monk, G. Mendel, discovered the unblending quality of genes around 150 years ago during his spare time while teaching at a monastery. We can say without an ounce of doubt that he did not have a first rate computer to work out his mathematics and necessary computations for his experiments.

Monasteries during the second half of the XlX century in Eastern Europe, Czech Republic were not known for the abundance of anything. However, writing paper and pencil or pen, he must have managed to keep in his dormitory room. The monk and geneticist Mendel knew that his discoveries (genes) were very important, but his fellowmen did not appreciate it. In the year 1865 he gave a lecture on his work at a local medical society followed publication of his discoveries in the society's journal. Unbelievable or not, his findings remained unknown to the scientific world and public in general until the first decade of the XX century. Mendel had proven scientifically what farmers had been practicing for thousands of years; they had been selectively taking the best and desirable offspring (gene traits) of animals, birds and plants to improve their stock. It appears to me that Mendel work is just recently appreciated during and after the sequencing of the human genome. We are learning to engineer genes for medical and food research for the benefit of humanity in general. A great section of our essay has been centered about DNA, genes and proteins. Gregor Mendel was born in 1822; he did not live to see the fruit of his work appreciated beyond the walls of his monastery. For me and many others that love his work, Mendel is another hero that deserves to be paraded along the Heroes Canyon on Broadway in Manhattan. Without insulting anyone, the science of genetics was born by counting round and wrinkle peas by a monk who was not known by his intellectual acumen.

My inquisitor is asking me to include G. Galileo, Marie Curie and Rosalind Franklin on the list of outstanding scientists to be selected for the Heroes Canyons. You should have a voice in the selection of our heroes. We did not mention Francois Jacob and Jacques Monod because they were paraded along the Arch de Triumph in Paris.

Mendel based his conclusion by observing hereditary determinants-genes- as phenotype, meaning their outside appearance or expression. With pea plants he observed and counted spherical and wrinkled peas. He began his work by selecting and observing pure spherical and pure wrinkled peas. He paired or fertilized a few of these so called pure breeds. The first generation of mated or fertilized peas, known to geneticists as F1, came out all spherical, the phenotype or external appearance of each pea he collected. Moreover, the scientist monk mated or fertilized the offspring of the first and subsequent generations of his selected peas and found out that they came out spherical and wrinkled peas in a ratio of 3 to 1. The spherical determined traits in peas he called dominant; and the wrinkled peas, the recessive hereditary determinants. What Mendel called hereditary determinants is known to us as genes.

Without tools to look inside each pea to examine and analyze how hereditary determinants work and remain "particulate", he had half opened the door into reading the book of life. Future scientists became interested in discovering not only how and where, are the hereditary determinants-genes-formed and stored? They wanted to know the alphabet or letters that would enable them to read how peas, flowers, plants, animals, and above all, how humans beings are being built from apparently, two simple molecules. We do not tired thanking that simple and humble monk for showing us, among other things, the outside expression of genes. At present time we have become engaged in an investigation race or hunt for all sorts of genes, including mutated genes, disease provoking genes, rare gene variants, comparative gene evolution among species, and many others. We wanted to identify, characterize and look inside genes and use its million years old code for our benefit. After a short hesitation whether hereditary information was stored in proteins, amino acids, or in similar chemical structure within the cell, scientific based experiment proved without any doubt that the genetic code resided in the DNA molecule.

Among the big players during this hunt for genes was L. Pauling, known to many scientists as the world leading chemist during his lifetime. Besides Pauling, there was Rosalind Franklin, Maurice Wilkins, and of course, our intellectual icons: J.D Watson and Francis Crick. We are not overlooking the contribution made by the world best known neurologist, Santiago R. Cajal and many others that would be impossible for us to mention here. Cajal made it clear for us that our brain is not a long sausage coiled around itself. He set it clear that our brain is composed of neurons and glia cells that are independent of each other, but communicate with each other through space between them known as synapse. They send and receive messages among themselves using chemicals called, neurotransmitters. For further information on the subject see the following books by this author: Parkinson, Alzheimer and schizophrenia as well as, The Brain, the world most wonderful organ. Returning to Watson and Crick double helix and, " it has not escaped our noticed that the specific pairing we have postulated...", pulled out biology from cataloging and classifying animals, plants and birds into a scientific race just rivaled by that of the atom. From that date on, Presidents and Prime Ministers on both sides of the Atlantic appeared on the front page of newspapers and scientific journals with Nobel Laureates and pioneers in biology.

Watson and Crick had set in motion a race among scientists around the world looking how amino acids were manufactured and built into proteins. The genetic information Mendel was referring to was found to be encoded in the sequence of nucleotides-the letters of DNA. However, a clear and accurate look at DNA and how it determines its nucleotides formation still needed some extra work and delicate investigation. At the core of this investigation and intellectual curiosity was none other than how genes work and behave at molecular level. Genes, proteins, diseases, drugs to cure it, and research tools seemed to have occupied the mind of scientists around the world. Out on the horizon, far away and hardly conceptualized by most people, there were scientists attempting to understand gene behavior and the sequencing of the human genome. Questions like, how do we look at, uncoil or unwrap DNA strands and the chromosome itself, were academic debate with very few scientists engaged in laboratory work. We wanted to know how long strands of nucleotides coiled up in beads like structures containing multiple amino acids, and still conserving, unblended, Mendel's particulate quality.

Ligaze Proteins

This biology race was turned into a DNA-RNA technology revolution. Biologists, chemists, physicists, mathematicians and computer engineers began to see the advantage of working together. Among the first DNA revolution technology tools we came across were restriction enzymes. It meant scientists could cut DNA molecules segments using the newly discovered restriction enzymes. The cut DNA segments, meaning a gene, or clusters of genes the scientists were interested in, were kept away, isolated for study, analysis and purposeful use. The next step scientists took upon themselves following restriction enzymes discovery, was making use of a wise protein, an enzyme dubbed ligaze. This ligaze enzyme could be used to paste or glue together the DNA sequences- genes-for whatever reason the researcher had in mind. According to J.D. Watson, there was a young scientist gifted with a business mind besides his enthusiasm for the genome sequence. His name, "Herb Boyer was already an expert on restriction enzymes in an era when hardly anyone had heard of them," wrote Watson. Well, at this early stage of DNA technology we had mastered two enzymes to do work for us. There were two enzymes we could use to cut and splice or joined together genes. Was it not a wonderful discovery! Mendel played with phenotype expression of genes; but we were touching and manipulating the genes themselves. Furthermore, the technology of cutting and gluing genes together became known as a plasmid. A plasmid is roughly like a CD for the recording and storage of information. By luck or coincidence, there was another brilliant scientist, Stanley Cohen, attending a Honolulu conference in 1972 on plasmids. So, by the grace of curiosity and scientific interest, "the first among plasmids" S. Cohen and Herb Boyer, the restriction enzymes star, realized that together, they had a DNA technology nobody else knew about.

Besides their interest in genes, their chemistry seems to have worked very well, and together, they established the first recombinant DNA company in the world. This was the world first genetic engineering establishment. All this play around with genes was done with small animals. The big job, the sequencing of the human genome was resting, if at all, on the drawing board of optimist biologists and scientists.

Boyer and Cohen's discoveries and DNA technology tools was not a small thing to go unnoticed by the scientific community. They could play-engineer- with bacteria and small animal cells, make cut with restriction enzymes and put it back together with a ligaze enzyme. Cutting and joining together compatible segments of genes gave birth to another novel idea, recombinant DNA technology. One dear parasite in our intestines, E Coli proved to be an excellent subject to work with. It is cheap and hardly demands much of you. How to transfer DNA material from one organism to another was the next challenge. Well, we can make use of plasmid to do the job. The question that lies ahead is how would the host receive the guest? You already know how one organism tends to defend itself and reject what it considers invaders or intruders. We are referring here to our defense system or immune system. There is a group of defensive cell that originates in the bone marrow, with T cells maturing in the thymus. It is a survival system that has defended you for thousands of years. It mobilizes thousands upon thousands of cells to keep you alive when an invader tries to break into your body. What I am saying is that the solution to recombinant DNA technology rejection problem already exists in your organism, particularly, in your brain. Boyer and Cohen had the solution at hand using a plasmid, meaning test tube experiments. Plasmid already cut in segments would glue together under a ligaze enzyme. At first glance it did not provide much help to Boyer and his plasmid pioneers. The ligaze enzyme was doing what was already known to the researchers, gluing stretches of plasmid together. However, the opportunity to come out of the laboratory with his test tube in hand crying out, Eureka, eureka, meaning, I found it, I found it, was not far away. During a following test, it was observed that the gluing enzyme would manipulate a cut plasmid to incorporate segments of DNA from neighboring plasmid, thus creating a hybrid. It was in Cohen hands to use his already proven technology to transplant the whole plasmid inside a bacterium for multiplication. Resistance to the newly added guest was overcome using already available antigen resistance. The next step would be the creation of a hybrid from two different species. We have an abundance of flies and birds we can capture and use in laboratories for the benefit of mankind. I remember crows coming to my fathers corn field to eat corn on the cob despite our strong objection. My father spent money and time planting and cultivating it, while crows had a feast at our expenses. We could capture a few crows, extract a few cells, cut DNA sequences using restriction enzymes, glue it to a bacteria plasmid using a DNA ligaze protein enzyme, and bingo; we came out with recombinant DNA molecule. It looks very simple the way we have described it in here. Perhaps after we practiced it a couple dozen times, it would be a simple job for all of us.

TV shows and DNA finger printing

Up to this point, most of the work had been done with bacteria; the next step would be with higher animals having similar genome like our little brother, the mouse. Recombinant DNA technology is and will be an excellent tool for the discovery of DNA mutations that are responsible for inherited diseases. It is already a first rate tool in medical research and therapeutic tools beginning with insulin for diabetics (Boyer and Cohen) as well as blood clotting proteins. Another tool we cannot ignore in this essay is DNA fingerprinting, although the finger is hardly involved in the game. Sometimes I enjoy television shows presenting two or more young fellows claiming or disclaiming paternity privileges or monetary contribution to the mother of a child. At times, I believe the TV show host presents two or more males with the same claim, is just a charade or stunt to arouse the spectator's feelings and improve its show rating. However, DNA technology has come out to be the jury and judge in such cases declaring who is the father of the child in question. DNA identity is 99.99 percent accurate despite the mother allegations that she slept only twice with the other young fellow. Besides paternity issue, I like to watch crime stories when DNA technology is involved in solving a crime.

Attempts to hide DNA finger print.

There have been criminal cases extremely difficult to solve for police officers and detectives to identify the criminal agent. The delinquent subject has been a cold and calculated person who has carefully considered all possible clues he might leave behind in the scene of the crime. I hear police officers and detective call him a professional criminal whose specialization is eliminating all traces of possible identification.

Among the things he might do is covering foot track in multiple ways, cover his face wearing mustache, paint or enlarging his eyebrows, modifying the shape of his nose , altering his voice, etc. He may appear to be the perfect serial killer, rapist or Bank robber. He is place on the list of most wanted criminals, and has hundreds of agents watching for a man with specific detailed description. He might change his phenotype or outside appearance, but his DNA profile is forever, no changes allowed. This criminal might have forgotten that his DNA left on a cup of coffee, soda or cigarette butt is enough to put him behind bars after he has seen the awaiting judge.

Leaving behind the serial rapist, bank robber and paternity claimants, we return to our wonder working enzymes. Restriction enzymes cut DNA molecules at specific sites. They know exactly where to go assisted by small segments of genes. Those cells of yours are not stupid at all; they have developed through the years many tricks for self-protection. We are just now learning about it. While our scientists were doing their job investigating the inside of bacteria, they observed that the host bacterium cut plasmid DNA segments at particular places. The mysterious enzyme (restriction enzymes) cut DNA only at certain nucleotide sequences.

They were very particular on choosing the sites to do the cutting. Moreover, there was another very interesting factor or piece of the game that had to be looked upon and included in the game. Not all bacteria possess the same restriction enzyme; therefore, some restriction enzymes would cut at different and selected sequence of nucleotide. One of the rules of the game practiced by bacteria cutting enzymes is that they choose to cut short, from four to eight nucleotides pairs. The following insight into cutting enzymes was that knowing that there are different bacteria, each species or subspecies would have its own enzyme, cutting at specific sequences. These discoveries on bacteria enzyme cutting game proved very useful for future sequencing of large genome. Since restriction enzymes would cut at certain sites containing different amount of base pairs, meaning that in a properly and adequately prepared aqueous medium or gel, the longest and heavier cuts of base pairs would go or assemble at different places from shortest segments of base pairs. It sounds logical and even practical to me that scientists thought that way at the time. Brain power plus dedication and perseverance make them a breed apart becoming Nobel Prize winners. The abundance of bacteria in our environment provided us the opportunity to have multiple restriction enzymes cutting at different sites available to us and ready for use in laboratories. Engineers in the field take among multiple restriction nuclease available, a specifically tagged, meaning a given restriction nuclease from the whole bunch available and ready for a job.

Electrophore Technology

These restriction enzymes will cut at a given DNA molecule at the specified site intended by the technologist. DNA fragment separation from each other is mostly accomplished using gel electrophore separating fragment or base pairs within fragments. One relatively easy and new way of doing it, is by staining DNA with a dye that fluoresces under ultraviolet light. (In the ocean beach in Lajas, Puerto Rico, when the Sun gives way to the night and the stars adorn the sky, you can see thousands upon thousands of tiny fish carrying a fluorescent protein separating them from any other fish. Take a short vacation and visit it; you may be motivated to do some type of research with those beautiful and interesting fish.) If a tiny fish could develop that kind of technology, your brain can do even better. Another model use by scientists is to incorporate radio-isotopes into DNA molecules. Cold Spring Harbor Laboratory on Long Island in New York is a pioneer center of research in this field. Our dear friend J.D. Watson made his first presentation of the double helix there.

Recombinant DNA

The DNA technology is using the most advanced equipment available to commercialize its products in many journals. With tools available to DNA technologist, long DNA molecules can be cut to the desired sizes fragments for the intended job, and join it together with the protein-enzyme we have already mentioned. The dogma by Watson and Crick that strands will attract each other to form the double helix based on the chemical attraction of nucleotide adenine to thymine and cytozine to guanine do not need you to worry about while playing with DNA fragments cutting and subsequent union. This DNA chemical ruling governing all organisms permits DNA from practically any animal source to be joined together. This chemical process allows technologists to have a free hand cutting and putting together DNA fragment from different organisms. The enzyme ligaze can join together two fragments without asking questions about their origin or place of birth. And, the energy to do the job is available and plentiful using another protein, ATP molecule to bring together the sugar-phosphate backbone. Backbone fragments are identified with digits 3 and 5 in most illustrations. The three end of a backbone has the shape of a half moon or the letter C while the 5 end is the full moon or a spherical tennis ball. When joined together they will fit perfectly into each other. You do not have to use a pair of pliers to do it; enzymes will do it for you. During this apparently simple process you are playing with chemical attraction of base pairs, energy providing ATP, restriction enzymes and DNA ligaze enzymes. When these processes are taking place inside your body, you would have supervisors and guardians proteins watching over the job in case there is an error. Your cells are always on super-alert status to intervene when it is necessary to do it. This all falls within the code of life your cells have kept for your own safety.

Laboratory workhorse, a bacterium

Base pairs in DNA segments are the same in all living organism, once you join it together as in plasmid, it will continue to replicate as if it were coming from an original and same mother. Basically, bacteria seem to be our best workhorse for duplicating our DNA fragments. Cloning is a process of making identical copies of a subject you want to duplicate. Our laboratory working horse, bacteria is the excellent tool to have DNA fragments copied several times fast and economically prudent. Every time a bacterium replicates or duplicates itself will also replicate the DNA fragment you inserted into it.

You may want to or need to replicate human DNA fragments making use of another organism for future use. You may need to cultivate DNA fragments from a person to repair an ear or nose. We would like to share with you that this is not science fiction. When working with DNA fragments introduced in a bacterium, it needs some protection from inside hostile fragments; otherwise, it will become food for the bacterium. A DNA enthusiast at the time (Stanley Cohen) was nearby with a plasmid to be used as a carrier, also known as vector. This carrier model,-a plasmid-possess a replication original which will duplicate independently of the bacterium chromosome. Manufacturing the final product, restriction enzymes and DNA ligaze play a major role on the lasso shape or circular plasmid in recombinant DNA. Recombinant DNA is copied as bacteria divides fast many times in our scientists laboratories.

A good trick in this very interesting game is to replicate one's own DNA fragment free of bacterium harmful content and keep it frozen for future use. The final step during recombinant DNA plasmid is the purification of the DNA from the content of the rest of the cell including the removal of the bacterium chromosome. The final purified plasmid DNA may contain thousands upon thousands copies of the original DNA fragment. There are sophisticated tools to get the DNA fragment clean and stainless out of the plasmid available to you. The thing that impresses me most is that all this new technology was and is stored in your brain cells.

Just one more time

Once more time we will continue to elaborate on DNA technology available for research at local individual laboratories. We had stopped at restriction enzymes, plasmid and recombinant DNA and at our icon pioneers: Boyer and Cohen for many obvious reasons unable to mention here. There are other equally outstanding pioneers in the field we did not elaborate on in order to concentrate on DNA and its multiple tools. However, we feel obliged to write something on another Grand Canyon hero, Warner Arber, a Swiss biochemist whose main concern at the time was, why some viral DNA were degraded after it was introduced into a bacteria. Those were days of grand pronouncements on the double helix, amino acids, triplets and protein synthesis. They were subjects of investigation conducive to Nobel Prize and big celebrations in luxurious hotels. Arber brain was busy at home wondering why some cells were prejudiced against some virus DNA. Host cells would selectively destroy some viral DNA. What characterized or distinguished aside that viral DNA from the rest of DNA if that great micromolecule of DNA is the same whether we are talking about micro-organism like virus, bacteria, an elephant or a sequoia tree in California. Arber sleep was interrupted wondering whether an unknown molecular, chemical process or bugging genie was inside the cell that kept its own DNA untouched; but fiercely attacked viral DNA. He faced the challenge with great enthusiasm and inquisitive mind and discovered that the genie in question was no other than the now very famous restriction enzyme that is first rate research tool in many DNA laboratories. It became the darling enzyme in recombinant DNA.

Restriction enzymes in bacteria cells restrict viral growth by cutting foreign DNA. This is the aha! insight that Werner Arber was waiting for, even when asleep. A lot of science work is done in your head proposing ideas and solutions to problems that end up in the laboratory, or rehearsed with peers and co-workers. Further investigation with restriction enzymes showed the DNA cutting is not random and wild job; but it is a sequence specific reaction. It does not cut capriciously and biased. A given or particular enzyme will cut DNA only when it recognizes a particular sequence.

You and I have to agree that it was not a small and insignificant discovery that could have remained hidden among multiple notes and rusty tools. This type of discovery is for you to stop biting on your sandwich, grab your cell phone and contact you friend immediately.

My noxious friend and tenant, E. Coli

We already covered for you Eco R1 as one of the first restriction enzymes to become a star among enzymes. It recognizes the sequence of the following six letters: GAATTC and cuts it from the rest of the letters. Up to now it sounds and looks easy and beautiful. Eco R1 saw the six letters sequence that began with a G followed by two AA and two TT ending with a C and made its cut. Stop and think, do not celebrate anything yet. How come Eco R1 does not cut the genome every time there appears the same sequence of letters? There would be a lot of pieces of genes and/ or fragments from the genome lying around strewn on the table, test tube, petri- dish and perhaps on the coffee pot. W. Arber was doing his science 24/7 and discovered that the bacterium that he was working with, manufactured another enzyme; this newly discovered enzyme did not allow automatic cutting of those six letters we named above. Eco R1 cutting privileges were restricted by another enzyme from the bacterium. This newly found enzyme modified the structure and chemistry of those six letters sequences of its own DNA. This modification of structure and chemistry was done by adding a methyl group to the bases. The methyl group consists of one carbon atom and three hydrogen atoms. The cell itself had invented the cutting restriction enzyme, the sequence cutting interval and the chemical modification to protect its own DNA being cut automatically. When Eco R1 was running around cutting prearranged sequences of letters; it did not recognize the six letters, otherwise this enzyme would make its cut without hesitation. It looks cool and easy, and should be. Bacteria have all these tools; we, as superior organisms have it and W. Arber proved it for us.

DNA sequence cutting was moving along quite well with new discoveries made on both sides of the Atlantic. DNA technology jargon was frequently heard in bars and restaurants and even in barber shops. The next step in our DNA technology festival we referred you to, there is a problem that scientists and medical doctors were aware of; it is bacteria resistance to antibiotics. Some people were responding negatively to antibiotics because the medication prescribed did not kill the bug inside the body of the sick person. The bacterium went through a mutation in its genome. The newly discovered antibiotic resistance was provoked by incorporating an alien or foreign small fragment of DNA later called a plasmid. A plasmid has the shape of a lasso or loop. Plasmids have made bacteria their main permanent place of residence. This lasso-loop- structure that lives within a bacterium is not a passive, inactive and uninvolved piece of DNA. It possesses the ability to replicate and pass on, along with the rest of bacterium genome, while the cell goes through its stage of division. The cell cannot leave it behind or get rid of it during cell division. We have repeated several times, most cell have survived millions upon millions of years. First we learned that bacterium, a cell, survived threats by mutation and secondly, cells have another survival tool, the plasmid. Surprises are not absent during investigating micro-organisms like bacteria. Plasmids were found to pass from a single bacterium to another which originally did not possessed it. It was a blessing for the receiving bacterium because it received DNA genes providing antibiotic resistance.

It remained for Stanley Cohen at Stanford University to further elucidate for us the new antibiotic tool, the plasmid. As the story goes, Cohen and Boyer met at a conference in Honolulu in Hawaii; both pioneers in restriction enzymes and plasmid gave birth to Recombinant DNA. One of the first products of this union was insulin. We enjoy sharing with you these important episodes of great scientific discoveries of our time. This stage of genetic engineering presented serious questions we had never encountered before. It meant we could play with genes at will cutting, adding and exchanging fragments of DNA among surprisingly different species. We manipulated genes from bacteria to bacteria discovering and forming antibiotic tools for the benefit of the many around the world. We are composed of DNA. There are frogs, mice, chickens, sheep, dogs and cats we could begin transferring genes- plasmid- for medical and research discoveries. It does not go unnoticed of the immense use of and benefit it could bring to the food industry, including vegetables seeds and roots. A second discovery of our planet earth was coming as scientists were exploring micro- organisms never seen or heard of before. A new scientific language has been invented to address new frontiers in all areas of modern science. It has opened doors for women to explore and succeed in this new field of study. We have traveled a long way from scientist Marie Curie to present time. Two of my icon neuro-scientists: Nancy C. Andreasen and Jeanette Norden are worldwide known professors whose teaching talents have motivated thousands of students. They are outstanding researchers and widely acclaimed University professors we love to listen to their lectures and read their books. During all these DNA technology discoveries; it was not a single Frankenstein working behind doors in his castle in far away, Transylvania. These were men and women exchanging information by phone, journals and conference all over the world. The molecular biologists, geneticists, chemists, paleontologists, physicists as well as farmers and fishermen were turning their attention to this new technology. Virus and bacteria were the objects of meticulous study for possible use in many areas of related investigation. Farmers the world over interested not only improving the output of crops; but also, antibiotic for their seeds and growing plants. Our hero, J.D. Watson was already head of the Cold Spring Harbor Laboratory on Long Island, a hot spot for enthusiastic and highly intelligent young scientists racing to make their discoveries first page news.

A Research Hiatus

E Coli was a very hot tool of research not only in the U.S.A. and Europe, but also in Asia. Dr. Watson was a cautious head scientist not in a rush to open a Pandora box. There was a five year hiatus of extreme cautious research; but logic and the mind of brilliant researchers closed ranks and went on doing first class research.

Going from micro-organism to four legged relatively large animals was no small jump. Inserting the desired small fragment of DNA into a plasmid to be transferred into a bacterium, provided for the bacteria to divide, multiply and grow producing a large number of copies of DNA fragments. This step might be called the first step involving the sequencing of larger genome. And..., among larger genomes, there was, in the mind of some visionary and enthusiastic scientists, the sequencing of the human genome. Once more time, the Atlantic Ocean was the divider between two brilliant pioneers in sequence technology. Wally Gilbert was at Harvard University and Fred Sanger was at Cambridge University in England. Sanger approach to sequencing using the same enzyme that copies DNA naturally in cells, the now famous molecule DNA polymerase, proved to be more practical for its use.

Protein enzymes functions

Protein and enzymes functions are many. They come in different shapes and are located on different organs and tissues of our body. Among their functions are responsibility for building necessary tissue for repairing occasional bruises and fighting and destroying undesirable and toxic bodily intruders like virus and bacteria. They have done a wonderful job throughout our evolutionary history healing fractures, wounds sustained while fighting and capturing animals for food. Enzymes functions are worthy of admiration, praise and scientific careful observation and investigation. There is a wise enzyme dubbed Dicer which in another essay I called it a horseman with a sharp sword. Among other functions, this horseman cuts long double stranded RNA molecules into shorter pieces. These short pieces are crucial for gene-silencing pathways that involve small RNA such as short interfering RNA or micro RNA which are the most abundant classes of small RNA molecules."[58] It means the cutting enzyme, Dicer, the same one I had called a horseman with a sharp sword is indispensable in generating the above named short RNA tools and their functions. A deficit of Dicer enzymes is believed to be involved in an eye disease- an advanced age -related macular degeneration. Previous work has shown that reduced Dicer levels can occur in many tissues and are associated with various diseases. Besides its crucial role in RNA- control gene silencing, mammalian Dicer1 (we belong to this group) seems to have another function. "It maintains visual health by degrading toxic RNA molecules." [59]

Scientist Hiroki Kanek, one of the leading investigators at the University of Kentucky at Lexington Wrote: "Our findings elucidate a critical cell survival function for Dicer 1 by functional silencing of toxic Alu transcripts. (Alu RNAs are

[58]- Bartel D.P., Cell 136 p. 215-233, 2009.
[59]-Nature, Vol. 471, 03/17/2011, p. 308

transcripts of Alu elements- the most abundant non- coding repetitive DNA sequences in the human genome.) This unexpected function suggests that RNAi- independent mechanism should be considered in interpreting the phenotype of system in which Dicer is deregulated. The author further alerts us by saying: "Recognition of Dicer1 unidentified function as an important controller of transcripts derived from the most abundant genome repetitive elements can illuminate new functions for RNases (enzymes) in cytoprotective surveillance." The horse with a sharp sword I wrote about six years ago is not only involved in diseases, but is actively engaged in the most abundant non-coding repetitive DNA sequences of our genome. You recall it is the same genome sequences we used to call "junk DNA" and some congressmen were complaining the Federal government was wasting money on junk waste. H. Kanek further cautious us writing, "To our knowledge, this is the first example of how Alu could cause a human disease via direct RNA cytotoxicity rather than by inducing chromosomal DNA re-arrangements or insertion mutation through retro-transposition which have been implicated in a thalassemia, colon cancer and neuro-fibromatosis." [60]

Following signals and functions of enzymes among pathways multiple cell interventions has proven an arduous and challenging task for researchers, but it is paying off our efforts in many ways. A few years back we were complaining about the excessive amount of non-essential glia cells occupying most of our skull; often referred to as a left over from our chimp heritage. Now we are learning they are involved in many unsuspected roles and games even necessary for normal brain function. The job of micro RNAs is not fully established yet despite its wide use as a research tool; but neither is the Dicer multiforms and functions. We can see they are involved in human diseases; but to identify the loci of a particular disease

---

[60]- Nature, Vol. 471, 03/17/2011, p. 325-329.

among billions of cells and trillions of synapses takes time and absolute dedication. Following RNAs during their multiple functions beginning with protein assembly to short interfering fragments cutting existing double strand and its possible participation in health and sickness demands a lot of attention from any researcher. When to stitch polymers together for a specialized job and when to do the cutting, even with gene fragment help, is a marvelous and ingenious intervention by any standard. Just for the sake of repetition, the job of manufacturing, degenerating and assembling units and repairing it at micro level when the need arises is a big challenge in my book of notes.   Gene splicing is used in recombinant DNA to insert and/ or to remove genetic sequences, those letters that we repeat over and over again, from one organism or sequences into another organism or another DNA sequence.

Learning the trade

How gene splicing took place led scientists explore and master to cut and splice nucleotides sequences, meaning recombinant DNA technology. Splicing or joining together any nucleotide sequence to build a new product for commercial purpose came under the new technology. Any genetic sequence can be cut using restriction enzymes and be spliced into another sequence of letters to manufacture a medical product for diseases that have remained undefeated until now. Restriction enzymes are used in laboratories to remove or to add- splicing-nucleotides to sequences. This technology has facilitated pharmaceutical companies to manufacture large amounts of essential proteins for medical purpose. Most diabetics have benefited from this wonderful DNA medical technology. We can insert insulin producing genes into the genome of bacteria to produce large amounts of proteins necessary for insulin genesis. Some individuals may call it cloning insulin producing genes.

We will make a final comment on this most significant scientific advancement during my lifetime; and perhaps during the last 2500 years ever since Socrates` death in Athens. During the processes discussed above, scientists discovered how their tool, the bacteria, had evolved to protect themselves from virus infection. A wise bacterium had learned to manufacture enzymes to recognize a specific DNA sequence on the invading virus and made a cut at that threatening sequence. There is an abundance of enzymes in nature that we need to know, meaning identify, that can cut the specific sequence of the intruder toxic invader. You just need the curiosity and enthusiasm of W. Arber to repeat his discoveries. The test tube in your laboratory is the ideal place to bring cells and extract the cutting enzymes you will need to play the game. The test tube is used to cut any DNA you want to work with. The DNA fragment that came out a cut by the enzyme can be used to splice- join together another DNA fragment or molecule- and bingo, you just created recombinant DNA.

If during your game of cutting and splicing together small and large fragments of DNA sequences, and the recipient of DNA you played with was a chromosome, you get another bingo; you can incorporate it- introduce it- into its host cell, thus receiving new genes. DNA fragments (genes) from any source can be transferred to any cell. We are all composed or made of DNA. It is the variations-mutations- within genes in the DNA molecule that made species grow differently.

There is no room for doubt if you follow the footstep of these great pioneers in DNA technology. There are many cutting- restriction endo- inside-nucleases in nature just waiting for you to begin a career in this field. Making cuts and splicing together all sorts of DNA will trap you forever. In my opinion, there is nothing more interesting and amusing than learning to play with your own book of life. You will learn how to engineer bacterium to use as antibiotic vector to save sick people and return them to healthy and happy life. You may become a genetic engineer working through worldwide agencies multiplying several times crop production by eliminating plagues, and saving millions of people from famine and subsequent death.

Viruses cannot multiply alone

You recall that virus by their own mechanism cannot make it. Virus, those nasty bugs need to infect the host cell by introducing their DNA. Viruses use the cell reproduction machinery of protein synthesis to multiply themselves. The virus becomes lord and master of the invaded cell. In a short while there will not be a thing left of the host cell; virus will be flying out looking for another victim to invade and destroy. During previous chapters of this essay you had learned how some cells have developed efficient tools for self-protection; and eventually, killing the toxic invader. We are grateful to dedicated scientists who pursued their research goal to its end. We have named a few for you. Their contribution to science, health and humanity in general, we have no words to thank them. There is a widely known and dedicated professor of Biology, Dr. David Sadava whose lectures I cannot ignore here. I love to hear him repeat things over and over again while taking real story from daily life episodes to illustrate and elucidate the message of his lesson. During one of his lectures, he described two ways or methods to get DNA into host cells. We will describe only one. You isolate a chromosome from a host organism and put it into a test tube; cut the chromosome with a restriction enzyme and the new DNA spliced- joined together- into it. This rec DNA is then put into the host cell, and since it is part of normal chromosome, the host cell thinks that it "belongs", and the rec DNA stays there and is replicated." [61]

---

[61]- David Sadava, Understanding Genetics: DNA, Genes, The Teaching Co. 2008.

Illness under the electron microscope

There are many illnesses whose etiology is just beginning to come out under the electron microscope and related medical research tools. Our immune system has been under the watchful eye of scientists for many years. There are scientists worldwide dedicated to find solution for our medical illness working 24/7. They are trying to unlock the door to our immune system, its failures and resulting diseases, among them lupus erythematosus. This illness is closely related to our autoimmune system. It produces or provokes chronic inflammation that may involve the heart, lungs, skin, joints, kidneys, central nervous system as well as circulatory system. Scientists in the field suspect that apoptosis is involved in the etiology of the disease. However, the abnormal production of antigens by B cells is claimed by outstanding researchers to be the primary culprit of this illness. Like schizophrenia, autism, bipolar, among others; it presents itself as a syndrome, making it more difficult to follow its pathways in billions of cells. We have seen young people almost crippled by the disease, unable to walk and limited to a wheel chair. Their joints, finger and knees with protrusion- swelling- making it even more difficult to hold fork and knife in their hands. This is followed by depression and occasional suicidal ideations and attempts. For a detailed description of B and T cells job in the immune system as it relates to lupus, see The Brain, the Wonder of the Universe, by this author, page 139 and on.

Anxiety

My first encounter with anxiety that I am aware of took place while attending Junior High School. I had a punishing teacher, Mr. Delgo, the object of one of my most recent books titled, An Episodic Toxic Memory, a survival story. I did not have to see him or hear his voice to have my heart beat so fast and loud that even my friends used to place their hands on my chest to compare it with theirs. He had punished and embarrassed me in front of the classroom that I began to experience very embarrassing episodes like wetting my pants, vomiting, diarrhea and headaches. Mr. Delgo, my teacher was an ogre occupying many neurons in my brain. I can say that in some respect he stopped me from growing up in many respect.

I had him for my teacher for three very long years. The numerous neurons and synapses involved in this episodic toxic memory were overly active-super-sensitized-firing and releasing chemicals, meaning, neurotransmitters- ultimately producing new dendrites sprouts with its consequent neuronal structural change. A few particular brain regions or cluster of cells were the primary recipient and reactive objects of Mr. Delgo abusive behavior towards me. Among those cells there is a group, a very old group of cells on both sides of the temporal lobe called amygdala. It is located in front-anterior- of another group of old brain cells named hippocampus. This primitive brain structure, to people in Greece and Rome who were curious and interested in brain anatomy, it looked like a seahorse, and so they named it.

My first investigation about fear and anxiety was centered on Sigmund Freud psycho analytic theory. My oldest brother was a fan of Freud during the late 1940 and early 1950. I recall I had begun to feel nausea, headaches, sleeping problems, nightmares besides a racing heart that occasionally I compared it with my father favorite horse, Canelo. I tried over the counter medication and herbal teas; my mother was believed to be an expert botanist among her immediate family and friends. In all honesty, neither my mother's expertise nor o.t.c. medication were of much help. My problem was not in my stomach despite my nausea and occasional vomiting. My problem resided inside my head somewhere in an outgrowth of synaptic connections I could not even imagine it existed at all.

Eric Kandel, Jeanette Norden and Nancy Andreasen, three of my earliest icons and intellectual stars, helped me get insight into the monster I had right inside my brain torturing me long after I had left Mr. Delgo behind. I had left behind Delgo in bones, blood and flesh, but his image, his bullying and all his wicked behavior was inside me. Later on, when I had moved to live in a great urban area with many experts in brain-mind-problems, I spent almost three years on a couch followed by another three and half years at a psychoanalytic oriented institute with significant modification from traditional Freudian theory. I must add that group therapy and group support were extremely helpful.

An almond looking cluster of brain cell

The almond looking brain structure we are referring to is the amygdala, that along with the hippocampus or seahorse, is involved in memory and learning. Without that almond looking structure positioned anterior to the hippocampus, we would not probably have made it in the jungle. I am not striking against animal living in the jungle, please. The new predator drives or rides expensive cars, wears expensive clothes, makes use of most modern cell phones and hires less sophisticated predators to do his job on street corners or expensive and selective bars.

I like to travel and my money has vanished from my pocket leaving me without a minimal of suspiciousness about the culprit or thief. You may argue that the amygdala is losing its usefulness, but perhaps it is the predator that has become smarter. The amygdala acts independently of nearby cluster of brain cell nucleus. You have the limbic system and prefrontal cortex that are emotion power centers; but the amygdala does not wait for feedback or response from them when it needs to put you in a safe place in a fraction of a second. Otherwise, you have become lunch or supper for a hungry predator. That amygdala of yours is often referred to as the brain door to emotions. There is a difference between fear and anxiety. In fear there is an object you are afraid of. In my case, I began to fear Mr. Delgo, later on, miles away, it was converted into anxiety. Anxiety has to be defined as a sustained level of elevated apprehension in the absence of an immediate threat. I had moved 1600 miles away from Delgo, my teacher, but the ogre in my brain kept torturing me day and night. He could not be an immediate threat to me at that distance, but he had already provoked a structural change on my neurons. When I saw a man who looked or spoke like him, anxiety overcame my apparent peaceful mind. When I was enjoying an ice cream sugar cone and a Mr. Delgo look like man passed by me, the ice cream turned sour and occasionally, made me throw up. Anxiety is considered by many clinicians as the most common psychiatric disorder. It may come from different sources and exist in all cultures and countries around the world. It may go unrecognized, misdiagnosed and untreated for many years in all levels of our society. Denial and the traditional stigma attached to individuals receiving help or treatment for brain problems provoke people to stay away from receiving help, and remain in hiding rather than getting professional help and regain their happiness. It goes without saying that alcoholism, drug abuse, marital discord and abuse have its base in anxiety.

Our subject of discussion, the amygdala has been implicated in anxiety etiology. Its implication is logical and is based on many laboratory experiments. Chronic stress can kill many amygdala cells, thus contributing to anxiety. Dead cells in the amygdala make you forgetful, learning is slower; you become irritated, blood pressure goes up, headaches are turned on and strokes and heart problem may also come to make you feel bad. Most medical doctors visit have an anxiety component. My story on the amygdala and anxiety is not intended to frightened you to run for help; it is best informative to familiarize you with relative minor problems that can be resolved before it reaches the brain and makes it his permanent home. "Despite the high prevalence of anxiety disorders in all levels of society, the underlying neural circuitry is incompletely understood according to Stanford University professor and researcher." Psychotherapy, among them Cognitive Behavior Therapy, Gestalt Psychology, Rational Emotive Therapy, and some goal specific and action oriented therapeutic modalities have been very helpful dealing with anxiety disorders. There are prescription drugs like the benzodiazepines, but their side effects and possible addicted component need to be seriously considered before you begin to take it.

In adding to incomplete studies on the amygdala, this investigation deficit can be partially attributed to the complex internal composition of the amygdala itself. The leading author at Stanford, Kay M. Tye wrote that the amygdala is composed of functionally and morphologically heterogeneous sub-nuclei with complex interconnectivity. The primary neurotransmitters are glutamate and GABA ergic medium spinal neurons. The centro-medial amygdala consisted of 95 percent GABA ergic while the basolateral amygdala is glutamatergic. No conclusive evidence of a specific causal factor or loci for general anxiety at the amygdala was found. Their investigation concluded that anxiety is continuously regulated by balanced antagonistic pathways within the amygdala. [62] During this essay as well as previous one, we have paid particular attention to the amygdala and the hippocampus not only because its antiquity in the brain; but also, because they are always on the hot spot. Both are primary survival cluster of brain cells that have helped this lad reach this point in life evolution throughout thousands of years. Soon, we will identify a particular protein in the amygdala responsible for breaking the established normal balance between glutamate and GABA production, thus provoking an emotional disorder. It is the amygdala continuous

---

[62] - Nature, Vol. 471, 03/17/2011, p. 358-362

exposure and over-excitation to stress that governs that particular protein. Likewise, we picked on translation processes where RNA is always a big player, but also, a hot spot for human illness. We should not close up and say, hasta la vista, until we pick another hot spot: cutting and splicing, particularly, splicing involved in many of our illness. Some researchers have begun to explore that sensitive area of cell manufacturing machinery. It would be wonderful for all of us if we could address this health issue with embryonic stem cell lines and re-generate a decaying amygdala.

Table showing the 21 amino acids and their codons

We will write them beginning with the largest number of codons to the smallest number of codons.

Arginine----------------AGA---AGG---CGA---CGC---CGG---CGU

Serine------------------UCA---AGU---AGC---UCC---UCG----UCU

Leucine----------------UUA---UUG---CUA---CUG---CUC---CUU

Threonine--------------ACA---ACC---ACG---ACU

Valine-------------------GUA---GUC---GUG---GUU

Glycine----------------GGA---GGC---GGU---GGG

Alanine----------------GCA---GCC---GCG---GCU

Proline------------------CCA---CCC---CCG---CCU

Isoleucine---------------AUA---AUC---AUU

Cysteine------------------UGC---UGU

Phenylalanine----------- UUC---UUU

Tyrosine------------------UAC---UAU

Histidine-----------------CAC---CAU

Glutamatic Acid---------GAA---GAG

Aspartic Acid------------GAC---GAU

Asparagine--------------AAC---AAU

Glutamine---------------CAA---CAG

Lysine--------------------AAA---AAG

Methionine--------------AUG

Tryptophan-------------UGG

Stop Codon--------------UAA----UAG----UGA

We arranged it from the largest to the smallest number of codons to code for protein synthesis responding to our curiosity and convenience. No relationship between each amino acid and its codons number from another amino acid was intended.

173

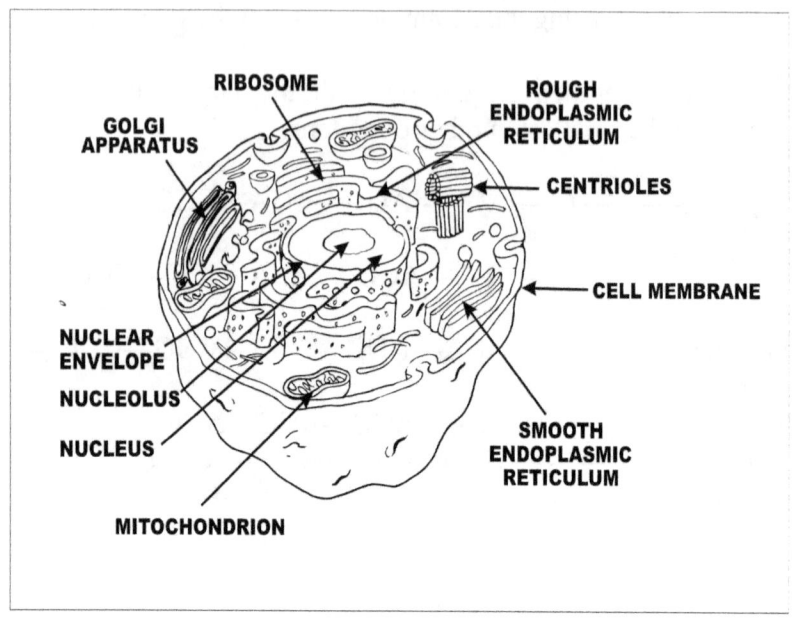

Inside an animal cell

There are dynamic chemical and molecular processes taking place inside each cell every fraction of a second of our life. Most of these processes involve extremely tiny units seen only under an electronic microscope. Equally small there are multiple vessels or organelles traveling back and forth inside the cell plasma carrying and delivering organic material necessary in our organism. You will find the endoplasmic reticulum, the Golgi apparatus, lysosomes and neighboring cells.

For instance, a tiny portion of the plasma membrane on the surface of the cell may separate and cut itself from the membrane, capture necessary neighboring subunits and bring it inside the cell for processing. Similarly, these tiny cut off portions may be used as carrier to nearby cells; there are always intercellular capture of material going on for cell processing and use. Captured external material is usually fused with lysosomes. These cellular vessels, the lysosomes, function as a de-generator machine digesting everything that passes by. Part of this dynamic process is importing and exporting cell material. The nucleus of the cell is surrounded and protected by a membrane often called, nucleus envelope. Right next to the nucleus, you will see the endoplasmic reticulum that is always busy processing and packing material inside the cell. Keep your eyes wide open because within this this particular cell vesicle you will find the protein assembly factory, the ribosome. In each eukaryote cell, meaning cells with a nucleus, there are many compartments with specific roles. The aqueous material-gel like substance- is called cytosol. Besides serving as a cushion, many cellular actions and reactions take place there. We must remind you that inside the cell there is the energy producing molecule, the mitochondria.

DNA Double Helix

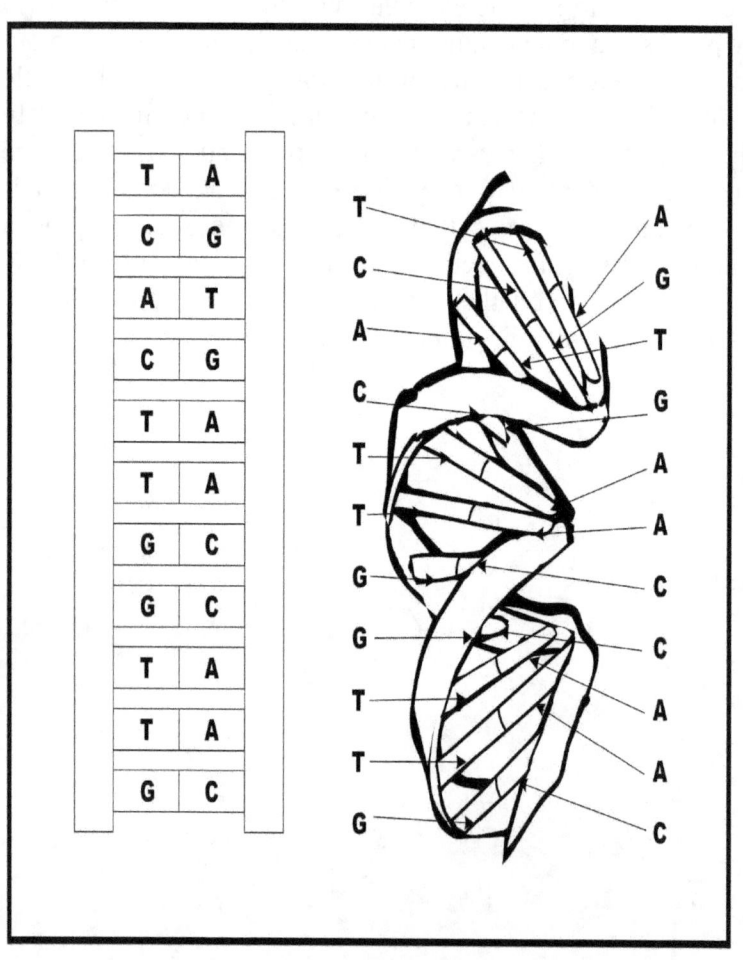

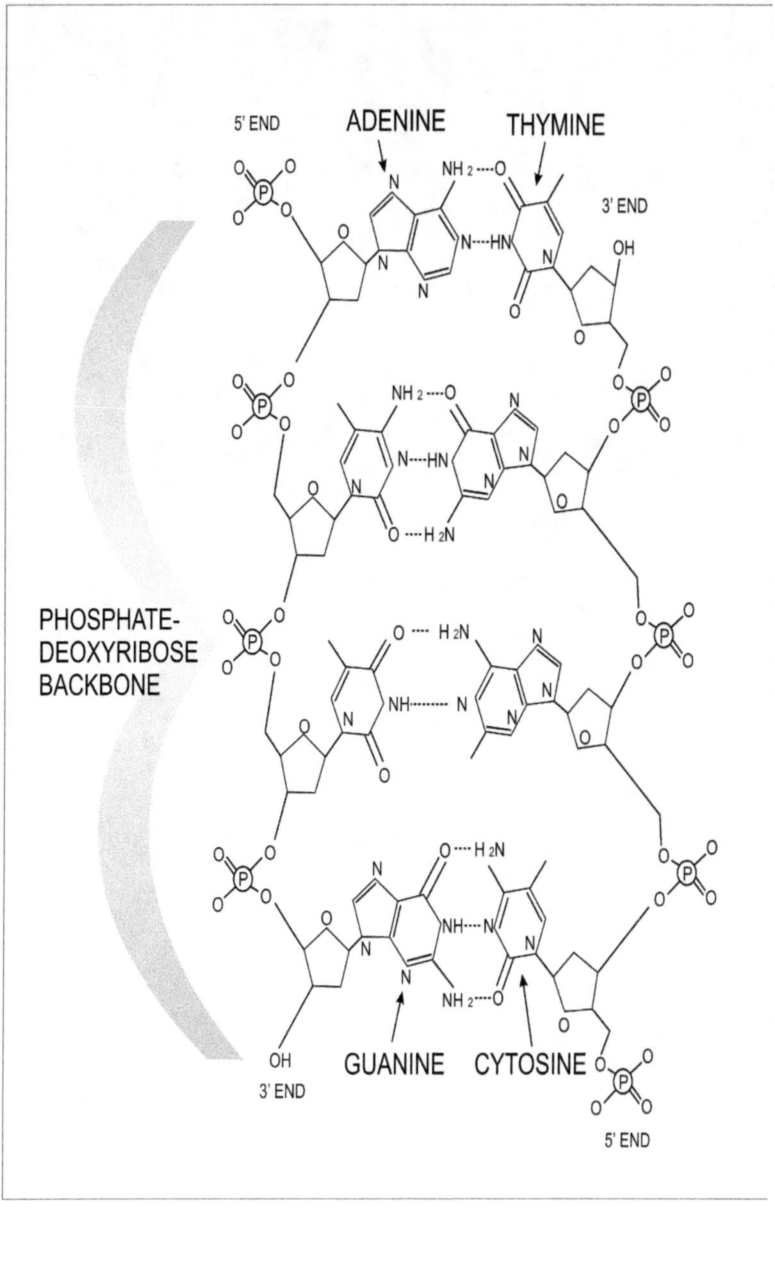

5' END   ADENINE   THYMINE

3' END

PHOSPHATE-
DEOXYRIBOSE
BACKBONE

GUANINE   CYTOSINE

3' END

5' END

177